JOSSEY-BASS TEACHER

Jossey-Bass Teacher provides educators with practical knowledge and tools to create a positive and lifelong impact on student learning. We offer classroom-tested and research-based teaching resources for a variety of grade levels and subject areas. Whether you are an aspiring, new, or veteran teacher, we want to help you make every teaching day your best.

From ready-to-use classroom activities to the latest teaching framework, our value-packed books provide insightful, practical, and comprehensive materials on the topics that matter most to K–12 teachers. We hope to become your trusted source for the best ideas from the most experienced and respected experts in the field.

Teaching the Common Core Math Standards with Hands-On Activities, Grades 6–8

Judith A. Muschla

Gary Robert Muschla

Erin Muschla

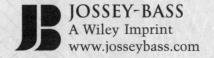
JOSSEY-BASS
A Wiley Imprint
www.josseybass.com

Published by Jossey-Bass

A Wiley Imprint

One Montgomery Street, Suite 1200, San Francisco, CA 94104-4594—www.josseybass.com

Jossey-Bass books and products are available through most bookstores. To contact Jossey-Bass directly call our Customer Care Department within the U.S. at 800-956-7739, outside the U.S. at 317-572-3986, or fax 317-572-4002.

Wiley publishes in a variety of print and electronic formats and by print-on-demand. Some material included with standard print versions of this book may not be included in e-books or in print-on-demand. If this book refers to media such as a CD or DVD that is not included in the version you purchased, you may download this material at **http://booksupport.wiley.com**. For more information about Wiley products, visit **www.wiley.com**.

Library of Congress Cataloging-in-Publication Data
Muschla, Judith A.
   Teaching the common core math standards with hands-on activities, grades 6-8 / Judith A. Muschla, Gary Robert Muschla, and Erin Muschla. — 1st ed.
      p. cm. — (Jossey-Bass teacher)
   Includes index.
   ISBN 978-1-118-10856-7 (pbk.)
   1.   Mathematics—Study and teaching (Middle school)—United States. 2.   Mathematics—Study and teaching (Middle school)—Activity programs—United States. 3.   Mathematics—Study and teaching (Middle school)—Standards—United States. 4.   Middle school education—Curricula—Standards—United States.   I. Muschla, Gary Robert. II. Muschla, Erin. III. Title.
   QA135.6.M87 2012
   510.71′273—dc23
                                                                                       2012001576

Printed in the United States of America

FIRST EDITION

PB Printing   10 9 8 7 6 5 4 3 2

ABOUT THIS BOOK

The Common Core State Standards Initiative for Mathematics identifies the concepts and skills that students should understand and be able to apply at their grade level. Mastery of these concepts and skills will enable them to move on to higher mathematics with competence and confidence.

Teaching the Common Core Math Standards with Hands-On Activities, Grades 6–8 offers activities that support your instruction of the Standards. The Table of Contents provides a list of the Standards and supporting activities and reproducibles, enabling you to easily find material for developing your lessons. The book is divided into three sections:

- Section 1: Standards and Activities for Grade 6

- Section 2: Standards and Activities for Grade 7

- Section 3: Standards and Activities for Grade 8

Each activity is prefaced with the Domain, which is a group of related Standards, and the specific Standard. For example, "Expressions and Equations: 6.EE.8" refers to the Domain, which is Expressions and Equations, Grade 6, and Standard 8. Background information on the topic for the teacher, materials—other than typical classroom supplies—and any special preparation that is needed are also included. Where applicable, the activities are identified with icons that indicate a major component of the activity will be cooperative learning ![icon], technology ![icon], or real-world focus ![icon]. All of the activities include a brief summary and specific steps for implementation. They are designed to build on concepts and skills that you have already taught and are ideal for expanding the scope of your instruction through reinforcement, enrichment, and extensions.

Each Standard is supported by at least one activity. The typical activity can be completed in one class period and focuses on application, demonstration of understanding, and communication about math. It may include creating mathematical models, charts, and graphs that explain concepts or skills; conducting investigations with manipulatives (both physical and virtual); playing mathematical games; writing problems and explanations; and making presentations that demonstrate students' understanding of a specific Standard. Because many of the activities offer multiple avenues for development and learning, we invite you to modify them according to the needs of your students. Answer keys are included where necessary; however, most activities are open-ended.

To augment your implementation of the activities, consider the following:

- Preview every Web site and work through any exercises so that you are better able to offer guidance during the activity.

- Place the URLs of Web sites in your browser to make the Web site easy to access.

- Enlarge and/or laminate any cards (often cut out from reproducibles) for future use with other students.

- Use a variety of instructional tools, such as traditional boards, whiteboards, overhead projectors, and digital projectors.

- For activities in which students play a game, you might want to provide a homework pass or other prize to the winners.

We hope that you and your students find the activities of this resource both interesting and enjoyable, and that the activities help you to guide your students to mastery of the concepts and skills of the Standards at your grade level. Please accept our best wishes for a wonderful year.

Judith A. Muschla
Gary Robert Muschla
Erin Muschla

ABOUT THE AUTHORS

Judith A. Muschla received her BA in Mathematics from Douglass College at Rutgers University and is certified to teach K–12. She taught mathematics in South River, New Jersey, for over twenty-five years at various levels at both South River High School and South River Middle School. As a team leader at the middle school, she wrote several math curriculums, coordinated interdisciplinary units, and conducted mathematics workshops for teachers and parents. She has also served as a member of the state Review Panel for New Jersey's Mathematics Core Curriculum Content Standards.

Together, Judith and Gary Muschla have coauthored several math books published by Jossey-Bass: *Hands-On Math Projects with Real-Life Applications, Grades 3–5* (2009); *The Math Teacher's Problem-a-Day, Grades 4–8* (2008); *Hands-On Math Projects with Real-Life Applications, Grades 6–12* (1996; second edition, 2006); *The Math Teacher's Book of Lists* (1995; second edition, 2005); *Math Games: 180 Reproducible Activities to Motivate, Excite, and Challenge Students, Grades 6–12* (2004); *Algebra Teacher's Activities Kit* (2003); *Math Smart! Over 220 Ready-to-Use Activities to Motivate and Challenge Students, Grades 6–12* (2002); *Geometry Teacher's Activities Kit* (2000); and *Math Starters! 5- to 10-Minute Activities to Make Kids Think, Grades 6–12* (1999).

Gary Robert Muschla received his BA and MAT from Trenton State College and taught in Spotswood, New Jersey, for more than twenty-five years at the elementary school level. He is a successful author and a member of the Authors Guild and the National Writers Association. In addition to math resources, he has written several resources for English and writing teachers, among them *Writing Workshop Survival Kit* (1993; second edition, 2005); *The Writing Teacher's Book of Lists* (1991; second edition, 2004); *Ready-to Use Reading Proficiency Lessons and Activities, 10th Grade Level* (2003); *Ready-to-Use Reading Proficiency Lessons and Activities, 8th Grade Level* (2002); *Ready-to-Use Reading Proficiency Lessons and Activities, 4th Grade Level* (2002); *Reading Workshop Survival Kit* (1997); and *English Teacher's Great Books Activities Kit* (1994), all published by Jossey-Bass.

Erin Muschla received her BS and MEd from The College of New Jersey. She is certified to teach grades K–8 with Mathematics Specialization in Grades 5–8. She currently teaches math at Monroe Township Middle School in Monroe, New Jersey, and has presented workshops for math teachers for the Association of Mathematics Teachers of New Jersey. She coauthored three books with Judith and Gary Muschla for Jossey-Bass: *The Algebra Teacher's Guide to Reteaching Essential Concepts and Skills* (2011), *The Elementary Teacher's Book of Lists* (2010), and the *Math Teacher's Survival Guide, Grades 5–12* (2010).

ACKNOWLEDGMENTS

We thank Jeff Gorman, Ed.D., Assistant Superintendent of Monroe Township Public Schools, Chari Chanley, Ed.S., Principal of Monroe Township Middle School, and James Higgins, Vice Principal of Monroe Township Middle School, for their support.

We also thank Kate Bradford, our editor at Jossey-Bass, for her guidance and suggestions in yet another book.

Special thanks to Diane Turso, our proofreader, for her efforts in helping us get this book into its final form.

Our thanks to our many colleagues who, over the years, have encouraged us in our work.

And, of course, we wish to acknowledge the many students we have had the satisfaction of teaching.

CONTENTS

Teaching the Common Core Math Standards with Hands-On Activities, Grades 6–8

Standards and Activities for Grade 6

Ratios and Proportional Relationships: 6.RP.1

"Understand ratio concepts and use ratio reasoning to solve problems."

> 1. "Understand the concept of a ratio and use ratio language to describe a ratio relationship between two quantities."

BACKGROUND

A ratio is a comparison of two numbers. A ratio can compare a part to a whole, a whole to a part, a part of a whole to another part, or a rate (a comparison of two different quantities).

Ratios may be expressed in three ways: with "to," by a colon, or by a fraction bar. For example, a ratio that compares the value of a quarter to a dollar can be expressed as 1 to 4, 1:4, or $\frac{1}{4}$.

 ACTIVITY: RATIOS ALL AROUND US

Working in groups of three or four, students will select a topic and write ratios that compare numbers associated with their topic. They will then create a poster, illustrating the meaning of select ratios.

MATERIALS

Math, science, and social studies texts; reference books, particularly almanacs and atlases; poster paper; markers; rulers; scissors; glue sticks. Optional: computers with Internet access.

PROCEDURE

1. Explain that numbers are constantly compared. Provide examples such as 1 inch on a map equals 50 miles, 1 pound of chopped meat makes 4 hamburgers, and a team's record of wins to losses is 2 to 1. These are all examples of ratios.

2. Explain that each group is to select a topic and write at least ten ratios associated with their topic. They are to then choose five of their ratios and create a poster that illustrates the meanings of these ratios.

3. To help your students get started, offer a broad list of topics and examples of possible ratios they may consider, such as:

   - Transportation: distance between cities, gas mileage, amount of luggage per person, costs per trip.

   - Cooking and baking: servings per person, cooking times per pound of food, ratios of ingredients.

   - Sports: per-game averages for individual players and teams, won-loss records, attendance.

   - Amusement parks: admission prices, types of attractions, roller-coaster statistics.

   - Information about their state, town or city, or school.

4. Encourage your students to brainstorm other possible topics.

5. After students have selected their topics, they should research the topics. Along with using books, they may also find the Internet helpful, especially for finding statistics on various topics. Remind them that they are to write at least ten ratios associated with their topics. From these they should select five that they will illustrate on a poster.

6. Encourage your students to be creative, neat, and accurate with their posters.

CLOSURE

Have each group of students present their poster to the class, explaining their selections of ratios. Display the posters in the room.

Ratios and Proportional Relationships: 6.RP.2

"Understand ratio concepts and use ratio reasoning to solve problems."

2. "Understand the concept of a unit rate a/b associated with a ratio $a : b$ with $b \neq 0$, and use rate language in the context of a ratio relationship."

BACKGROUND

A ratio is a rate that compares two quantities. A unit rate compares a quantity to 1.

To find a unit rate, divide the numerator and denominator of a ratio expressed as $\frac{a}{b}$ by b where $b \neq 0$. For example, if Milo ran 1 lap in $1\frac{1}{4}$ minutes, the ratio of laps to minutes is $\frac{1}{1\frac{1}{4}}$. The unit rate is found by dividing the numerator and denominator by $1\frac{1}{4}$ so the denominator is equal to 1. Milo ran $\frac{4}{5}$ of a lap in 1 minute.

 ACTIVITY: UNIT RATE TIC-TAC-TOE

This activity is best implemented in two days. Students will first work individually and then in groups of four or five, with each group divided into two teams. Each team will create problems about finding unit rates that the other team will solve. As they solve, or fail to solve, the problems, the teams will complete a tic-tac-toe board.

PROCEDURE

1. On the first day of the activity, explain that each student is to create five unit rate problems. They should make an answer key for their problems on the back of the sheet. Caution them not to show their problems to other students. After students have finished their problems, collect them and check that the answers to the problems are correct.

2. The next day, return the problems to their owners and divide students into groups.

3. Within each group, students should form two teams: One team will be the "X" team and the other will be the "O" team. One student should draw a tic-tac-toe board.

4. Explain the rules of the game:

- A member of the X team will read a unit rate problem, which the members of the O team must solve. If the O team solves the problem correctly, they may place an O on any square of the tic-tac-toe board. If their answer is incorrect, they must place an X on any square of the board.

- A member of the O team now reads a unit rate problem, which the members of the X team must solve. If the X team solves the problem correctly, they may place an X on any square of the tic-tac-toe board. If their answer is incorrect, they must place an O on any square of the board.

- The object of the game, of course, is to get three Xs or three Os in a row. The game continues until there is a winner or there is a draw. If a second game is played, a member of the O team reads the first problem. To play more games, students may need to create more unit rate problems.

5. As students play, you may find it necessary to assume the role of referee and provide an explanation for answers that students challenge.

CLOSURE

Announce the winners of the games. Review any problems that proved to be particularly troublesome. Ask students to summarize how to find a unit rate

Ratios and Proportional Relationships: 6.RP.3

"Understand ratio concepts and use ratio reasoning to solve problems."

3. "Use ratio and rate reasoning to solve real-world and mathematical problems, e.g., by reasoning about tables of equivalent ratios, tape diagrams, double number line diagrams, or equations.

   a. "Make tables of equivalent ratios relating quantities with whole-number measurements, find missing values in the tables, and plot the pairs of values on the coordinate plane. Use tables to compare ratios.

   b. "Solve unit rate problems including those involving unit pricing and constant speed.

   c. "Find a percent of a quantity as a rate per 100; solve problems involving finding the whole, given a part and the percent.

   d. "Use ratio reasoning to convert measurement units; manipulate and transform units appropriately when multiplying or dividing quantities."

BACKGROUND

Problems involving ratios can be solved using various methods, such as creating tables, tape diagrams, double number line diagrams, graphs, or writing equations. Most students are familiar with using tables, graphs, and equations.

To construct a table, use the ratio in the problem and create equivalent ratios. For example, if Sarah earns $200 during a typical 8-hour workday, create a table to show the amount of money she earns each hour. First, scale the ratio $\frac{\$200}{8hr}$ down to its unit rate by dividing the numerator and denominator by 8. The unit rate is $25 per hour. Using this information, you can create a table showing the money Sarah earns each hour. The data display in a table can also be used to find the missing values. For example, if Sarah worked 5 hours, write a proportion $\frac{4}{100} = \frac{5}{x}$ and solve for x. Sarah earned $125. If she earned $175, set up a proportion $\frac{6}{150} = \frac{x}{175}$ and solve for x. Sarah worked 7 hours. (Note that any ratio of time to money may be used in writing a proportion.)

Time (hours)	Money Earned
1	$25
2	$50
3	$75
4	$100
5	
6	$150
	$175
8	$200

The data from this table can also be used to create a graph. Time in this situation is the independent variable (graphed on the *x*-axis) and money earned is the dependent variable (graphed on the *y*-axis). The time and money earned represent an ordered pair on the graph. On this graph, the first ordered pair is (1, 25). An equation to represent this situation can be written from the table, graph, or unit rate. The equation $y = 25x$, where *y* represents the money earned and *x* represents the amount of time, shows the relationship between time and money earned. This equation can be used to find the amount of money earned or hours worked.

ACTIVITY: THE FASTER RATE

Working in groups of three or four, students will walk or run 20 meters and record their times. They will use their data to construct a table and a graph and write an equation. They will then use each representation to compare their rate to the rates of other students in the class.

MATERIALS

One stopwatch for each group; graph paper; at least two cones.

PROCEDURE

1. Take your students outside or to the gym and place two cones 20 meters apart. (If you have a large class, you may want to set up additional pairs of cones.)

2. Explain that students will walk or run 20 meters and record their data. As one student walks or runs, another student will use a stopwatch to track the time it takes the student to travel from one cone to the other. A third student will write this time down on a sheet of paper. The students will then switch tasks until each person has walked or run from one cone to the next.

3. After all students have finished walking or running, return to the classroom and instruct groups to write their rates as ratios of distance to time. Explain that they must create a table showing the distance they were at during each second of their walk or run, assuming they were walking or running at a constant speed. In order to do this, students must find the unit rate by dividing both the distance and time by the time. Instruct your students to round their answers to the nearest tenth. After they create a table, they must use the information on the table to create a graph of their data. Finally, they will write an equation, expressing their rate as a function of time.

4. After all of the students have represented their rates in a table, graph, and equation, instruct them to compare their rates with the rates of another group. In these larger groups, students should examine the tables, graphs, and equations to determine who was walking or running at the fastest rate. For each representation, they should discuss how the fastest rate is shown. For example, by looking at a table, the fastest rate is shown by the greatest increase in distance per second. In a graph, the fastest rate is

shown by the steepest line. In an equation, the fastest rate is shown by the slope. (The greatest coefficient of x indicates the fastest rate.)

5. Working as a class, list the results of each group on the board. Ask for volunteers to explain how they could use this data to find a percent of a quantity as a rate per hundred. For example, if a student walked 3 meters in the first second, find the percent of the distance he traveled. This can be expressed as $\frac{3\,m}{20\,m} = \frac{x}{100}$ or 15%. The student finished 15% of his total walk in one second. Note that students can verify this by finding 15% of 20 meters to get 3 meters.

6. Instruct your students to work in their original groups to generate a problem involving the whole, given a part and a percent. For example, suppose a student walked 15 meters, which was 20% of the total distance. To find the total distance, write and solve a proportion $\frac{20}{100} = \frac{15}{x}$. Share the problems the groups generate with the class. Students should solve the problems of other groups.

7. Working as a class again, instruct your students to use ratios to convert measurement units; for example, meters to centimeters, meters to kilometers, or meters to feet. For example, suppose a student walked 3 meters in one second. Meters per second can be converted to centimeters per second by using ratios in a proportion: $\frac{3\,m}{1\,sec} \times \frac{100\,cm}{1\,m} = \frac{300\,cm}{1\,sec} = 300$ cm per second. Note that setting up the ratios and proportions correctly ensures that meters cancel out and the answer will be centimeters per second. Also explain that because 100 cm = 1 m, multiplying by $\frac{100\,cm}{1\,m}$ is the same as multiplying by 1.

8. Instruct your students to work in their original groups to generate a problem involving the use of ratios to convert measurement units. Share the problems the groups generate with the class. Students should solve the problems of other groups.

CLOSURE

Discuss the overall results of your students as a class. Review the process of taking a rate and representing it in a table, graph, and equation. Also discuss how to determine faster rates in tables, graphs, and equations, and how students may use ratio reasoning to solve problems.

The Number System: 6.NS.1

"Apply and extend previous understandings of multiplication and division to divide fractions by fractions."

> 1. "Interpret and compute quotients of fractions, and solve word problems involving division of fractions by fractions, e.g., by using visual fraction models and equations to represent the problem."

BACKGROUND

A division problems has three components—a dividend, a divisor, and a quotient. For example, in $\frac{1}{2} \div \frac{1}{6} = 3$, $\frac{1}{2}$ is the dividend, $\frac{1}{6}$ is the divisor, and 3 is the quotient.

To write a related multiplication problem, multiply the divisor by the quotient, which will equal the dividend. $\frac{1}{6} \times 3 = \frac{1}{2}$. This may also be written as $\frac{1}{6}$ of $3 = \frac{1}{2}$.

 ACTIVITY: MODELING DIVISION OF FRACTIONS

This two-day activity requires students to work in pairs or groups of three. On the first day, each group will write a word problem, create a model, and write an equation to solve the problem. On the second day, groups will present their work to the class.

MATERIALS

Fraction bars and fraction circles to make models (rulers and compasses may be used instead).

PREPARATION

At the end of the first day, collect each group's work. Make a copy of the work that contains each group's problem, model, and equation. (You will return the copy of their work to the group to use as an answer key.) Keep the original and cross out the names of the group's members. Make enough copies of each original for each group to have one copy of every group's work. Use a paper cutter to cut each copy you made into thirds—each third containing a problem, model, and equation—which you will distribute on the second day. Each pair or group of students will then have a separate copy of every other group's problem, model, and equation.

1. On day one, explain that division involves finding the number of parts in the number to be divided. Be sure your students understand these terms: dividend, divisor, and quotient.

2. Offer this problem: How many quarters are in half of a dollar?

   - Present the following models to your students. The first one uses fraction bars to show the number of quarters in a half dollar, and the second uses fraction circles.

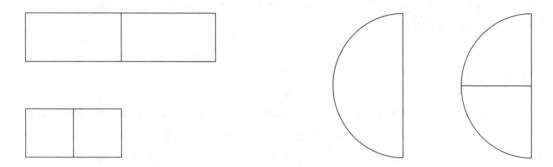

   - Ask your students how many fourths can be placed on the model of half of the fraction bar. Or, using the fraction circles, ask how many fourths of a circle can be placed in half of the circle. In both cases, the answer is 2.

   - Explain that this problem can also be modeled by the equation $\frac{1}{2} \div \frac{1}{4} = 2$. The answer can be checked by multiplying the divisor by the quotient, which should equal the dividend. $\frac{1}{4} \times 2 = \frac{1}{2}$, which is the same as $\frac{1}{4}$ of $2 = \frac{1}{2}$.

3. Explain that each pair or group of students is to create a word problem that can be solved by dividing fractions. They are to create a model using fraction bars or fraction circles that represent their problem and write an equation that can be used to solve the problem. Remind students that their work must be accurate and neat, because it will be read by other students. After everyone in the group agrees that their work is correct, they should recopy it in the following manner:

   - Start by holding a sheet of lined paper so that the lines are horizontal.

   - Divide the paper into thirds by folding the bottom of the paper up and the top of the paper down.

   - Write the problem in the upper third of the paper, sketch the model in the middle third, and write the equation in the lower third.

   After they have finished, each group should hand their paper to you, which you will then copy and cut into thirds. Organize the thirds into three sets, one set for the problems, the second for the models, and the third for the equations.

4. On the second day, randomly distribute the copies of the problems, models, and equations (all of which were previously cut into thirds) to each group. Each group should receive a problem, model, and equation that was created by every other group.

5. Instruct your students to match each problem with the model and its equation.

CLOSURE

Share students' results by having one member of a group read the group's original problem, model, and equation. The other groups can then check their work.

The Number System: 6.NS.2

"Compute fluently with multi-digit numbers and find common factors and multiples."

> 2. "Fluently divide multi-digit numbers using the standard algorithm."

BACKGROUND

The standard algorithm for long division involves the following steps:

1. Divide.

2. Multiply.

3. Subtract.

4. Check that the remainder is smaller than the divisor. If the remainder is larger than the divisor, increase the quotient by 1 and redo the multiplication and subtraction.

5. Bring down the next number and repeat the process until there are no more digits to bring down.

 ACTIVITY 1: SNORK'S LONG DIVISION

Students will practice long division on a Web site.

MATERIALS

Computers with Internet access.

PROCEDURE

1. Review the process of long division. Present examples such as $41\overline{)515}$ (Answer = 12 R23) and $58\overline{)3,604}$ (Answer = 62 R8) by explaining each step. Remind students that they can check the answer of any division problem by multiplying the quotient by the divisor and adding any remainder. If their quotient is correct, their answer should equal the dividend.

2. Instruct your students to go to the Web site http://www.kidsnumbers.com/long-division.php. Direct them to the box of Snork's Long Division and tell them to type in the highest number you want them to work with. They should click on "Play," and they will

be guided through the process of long division. After completing a problem, they will be given the opportunity to try another one by clicking on "Again."

3. Allow students about ten to fifteen minutes to practice problems on the Web site.

CLOSURE

Ask students to summarize the steps for long division.

ACTIVITY 2: LONG DIVISION RELAY RACE

Sitting at their desks in rows, students will complete a long division relay race. The first row to complete the most problems correctly is the winner.

MATERIALS

Blank sheets of paper.

PREPARATION

On a blank sheet of paper, write a long division problem on the upper left side. Write the quotient on an answer sheet for your reference. With a ruler, draw a line down the middle of the paper. At the top of the right side, write "Work Space." Do this four more times, writing a new problem on a new sheet of paper each time. Make photocopies of each sheet so that there is one set of sheets (five separate problems) per row. Label each sheet in each set by row, for example, "Row 1," "Row 2," and so on. (This will make it easy to keep track of correct answers for each row as the race goes on.) Place each set of sheets for each row in a separate pile. These sheets will serve for one relay race.

PROCEDURE

1. Arrange students' desks into rows, designating the rows by numbers.

2. Explain that students will take part in a long division relay race. Following are the rules of the relay:

   • You will hand a sheet of paper, face down, with the same division problem to the first student in each row. Note that students should work the problem out on the left-hand side of the paper, but may use the right-hand side for additional computation, if necessary.

- Once the race begins—on your signal—the first person in each row will do the first step of the problem, which is to divide. He will then pass the paper to the student behind him who will do the next step, multiplication. This student then passes the sheet to the student behind him, who will subtract. This student then passes the paper to the student behind him, who will compare the difference with the divisor, making any corrections, if necessary. He then passes the paper to the student behind him who brings the next number down and divides. This procedure continues until the problem is completed, at which point the student who completes the problem brings it to you. If a row has fewer students than other rows, the last person must bring the problem back to the first person in the row to complete the next step.

- Emphasize that at any step along the way, if a student finds a mistake made by someone earlier in the row, she should correct it. Otherwise, her row will not find a correct answer.

- After a row finishes the problem and the last person who worked on the problem brings it to you, you will give this student another sheet and the relay race continues by this student starting the division process on the new problem. In the meantime, you will assign one point to the row if their answer was correct, but no point if it was incorrect.

- The race ends when all of the rows have completed all of the problems. The row that completes the most problems correctly is the winner of the race. In case of a tie, the row that was first to finish is the winner.

3. Start the race. May the fastest (and most correct!) row win.

CLOSURE

Discuss any problems that students had difficulty solving.

The Number System: 6.NS.3

"Compute fluently with multi-digit numbers and find common factors and multiples."

> 3. "Fluently add, subtract, multiply, and divide multi-digit decimals using the standard algorithm for each operation."

BACKGROUND

Even in this age of lightning-fast calculators, understanding basic operations is an essential skill. Not only does a solid understanding of operations with decimals provide a foundation for mastering higher mathematical skills, but there will be times students will need to add, subtract, multiply, and divide decimals without the aid of a calculator.

ACTIVITY: DECIMAL OPERATION TOURNAMENT

Students will compete in a decimal operation tournament in which they will have to add, subtract, multiply, and divide decimals on the board. The fastest and most accurate problem-solver wins.

MATERIALS

Depending on whether you have a traditional chalkboard or whiteboard, you will need enough chalk or markers for three to five students to work at the board at a time.

PREPARATION

Make a list of problems—addition, subtraction, multiplication, and division—and their answers for the tournament. You might simply use problems from your text. Your list should contain about 30 problems. You should also have a list of students' names to use as a score sheet.

PROCEDURE

1. Explain the rules of the tournament:

   • You will choose students (the number will be determined by the amount of board space you have) to go to the board. Starting with one row and then working through the others is a good plan, making it easy to keep track of students.

- You will present a problem; for example, 0.4 + 1.23. Students will write the problem on the board.

- On your word, "Go," students will begin solving the problem. The first student to solve the problem correctly receives a point and remains at the board. The others return to their seats and are replaced by new contestants. Be sure to mark a point by the winner's name on your score sheet.

- You present another problem and the tournament continues.

- Students at their seats should also try to solve the problem. As they finish the problem, the first student to finish should raise her hand and call out—but not too loudly—"One," the second student "Two," and the third "Three." Only the first three should call out. Should no one at the board solve the problem correctly, the first of the students at their seats to raise her hand and solve the problem correctly wins the point. This student then automatically goes to the board with the next set of students. (She does not have to wait for her turn to come later.) If the first student at her seat does not have the correct answer, the student who called "Two" gets a chance to provide the answer, and if this student's answer is also incorrect, the student who called "Three" gets a chance. If no one gets the right answer, a new set of students replaces those who were at the board and the tournament continues. Students at their seats will quickly realize the advantage of trying to solve the problem, which is that they will get to the board more often where they will have a chance of accumulating more points.

- The game should continue until all students have a chance to get to the board a few times.

2. Start the tournament. Beware: your students will likely participate with enthusiasm and energy.

CLOSURE

Discuss any problems that the class found difficult to solve.

The Number System: 6.NS.4

"Compute fluently with multi-digit numbers and find common factors and multiples."

4. "Find the greatest common factor of two whole numbers less than or equal to 100 and the least common multiple of two whole numbers less than or equal to 12. Use the distributive property to express a sum of two whole numbers 1–100 with a common factor as a multiple of a sum of two whole numbers with no common factor."

BACKGROUND

A factor is a number that evenly divides into a larger number. The greatest common factor (GCF) is the largest number that divides into two or more numbers evenly. Students can find the GCF by listing the factors of two or more numbers and selecting the largest number that appears in both lists.

A multiple of a number is the product of the number and a counting number. A common multiple is a number that two or more numbers evenly divide into. The least common multiple (LCM) is the smallest multiple of the numbers. Students can find the LCM by listing the multiples of the numbers and selecting the smallest number that is on both lists, or they can find the product of the two numbers and divide it by the GCF.

The distributive property states that $a(b + c) = ab + ac$ and $(b + c)a = ba + ca$. It relates multiplication to addition because students can distribute a factor to each term enclosed in the grouping symbol.

 ## ACTIVITY: THE NUMBERS GAME

Students will work in groups of three or four for this activity. They will play a game in which they will find the greatest common factor, the least common multiple, and use the distributive property to express a sum of two whole numbers with a common factor as a multiple of a sum of two whole numbers with no common factors other than 1.

MATERIALS

3″ × 5″ index cards; dark marker; small whiteboards. Optional: graphing calculator.

PREPARATION

To play the game, you will need to generate sets of random numbers. Using a dark marker, write the numbers 1–100 on index cards, one number per card. Write the numbers 1–12 on a second set of index cards, one number per card. On ten index cards, write the sum of two whole numbers that have a common factor, one sum per card; for example, $12 + 15$. Base the sums on the abilities of your students. The first set of cards will be used to find the GCF, the second set will be used to find the LCM, and the third will be used with the distributive property. (Note: A graphing calculator may instead be used for generating random numbers, using the random integer menu.)

PROCEDURE

1. Write each group's number on the board or screen for keeping score; for example, Group 1, Group 2, and so on.

2. Explain how the game is played. Groups play against each other. The game has three rounds. In Round 1, you will present two random numbers. The students in each group are to find the GCF of the numbers. In Round 2, you will present two random numbers. Groups are to find the LCM of the numbers. In Round 3, you will present a sum of two numbers. Groups are to use the distributive property to rewrite the sum as a multiple of a sum of two numbers that has no common factor other than 1. Each correct answer is worth one point. The group with the most points at the end of the game wins. Play a sample of each round of the game so that everyone understands what to do. If necessary, review finding the GCF, finding the LCM, and using the distributive property.

3. Play the game as described below:

   - Start Round 1 by shuffling your first set of cards and randomly drawing two cards. Write the numbers of the cards on the board. Groups are to find the GCF of the two numbers. As students are working to find the answer, you should find the answer, too. Placing a time limit of 30 to 45 seconds for groups to work out their answers keeps the game moving. Each group should write their answer on their whiteboard and place the board face down. (Instead of the whiteboards, you may substitute clear page protectors. Place a white sheet of paper behind the page protector for a background.) Ask that a member of each group hold up the group's whiteboard. Check the answers and place a checkmark by the names of the groups that have the correct answer. Repeat the procedure. Choosing five to ten pairs of numbers makes for a reasonable round.

   - For Round 2, shuffle your second set of cards and choose two cards. Students are to find the LCM. Follow the procedure established in Round 1.

   - For Round 3, select a card from your third set. Students are to use the distributive property to rewrite the sum of the numbers as a multiple of the sum of two whole numbers that have no common factor other than 1. Follow the established procedure.

CLOSURE

Tally the scores. The group with the most points wins. For a tie-breaker, present another problem. The first group to find the correct answer wins.

The Number System: 6.NS.5

"Apply and extend previous understandings of numbers to the system of rational numbers."

> 5. "Understand that positive and negative numbers are used together to describe quantities having opposite directions or values (e.g., temperature above/below zero, elevation above/below sea level, credits/debits, positive/negative electric charge); use positive and negative numbers to represent quantities in real-world contexts, explaining the meaning of 0 in each situation."

BACKGROUND

Numbers are used to measure or describe real-world situations such as those noted above. Every number except zero has an opposite. 5 is the opposite of -5, $-\frac{2}{3}$ is the opposite of $\frac{2}{3}$, a gain of 1 pound, which can be represented by $+1$ or 1, is the opposite of the loss of a pound, which can be represented by -1. The meaning of zero depends on the situation.

In the set of real numbers, zero is the reference point that separates positive numbers from negative numbers. In terms of losing or gaining weight, for example, zero represents the original weight before any losses or gains.

 ACTIVITY: FINDING THE OPPOSITE

Working individually or in pairs, students will create an origami fortune teller whose flaps will reveal the opposite of a number or quantity, instead of a fortune.

MATERIALS

$8\frac{1}{2}'' \times 11''$ sheets of unlined paper, one per student.

PREPARATION

For each student, make an $8\frac{1}{2}'' \times 8\frac{1}{2}''$ square by cutting a $2\frac{1}{2}'' \times 8\frac{1}{2}''$ strip from a sheet of unlined paper. Go to the Web site http://www.dltk-kids.com/world/japan/mfortune-teller.htm where you will find directions for making an origami fortune teller. (You may also search for other directions with the term "directions for making an origami fortune teller.") Create an origami fortune teller for yourself. Not only will you have an example to show your students, but you will familiarize yourself with the procedure so that you can assist students in making their own origami fortune tellers.

1. Explain to your students that they will make an origami fortune teller, but instead of the flaps revealing a fortune, they will reveal the opposite of the situation described on the flap.

2. Direct your students to the Web site given under Preparation for directions. (You may instead prefer to project the Web site onto a screen and guide your students through the steps.)

3. As students are working on their origami fortune tellers, circulate around the room, providing help as necessary. There will be questions.

4. After students have completed folding their fortune tellers, instruct them to write on the flaps in the following manner:

   • Fold the paper in half and write "Left" on one square and "Up" on the other.

   • Turn the paper over and write "Right," then "Down."

   • Open the paper and turn it over to reveal eight triangles, as shown below.

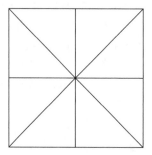

   • Choose one triangle and write a situation that can be modeled by a positive or negative integer; for example, 15° below zero. Do this for each of the triangles so that eight situations are described.

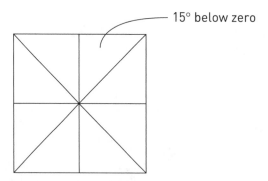

15° below zero

- Lift up a flap. Write the opposite of the situation described on the flap above it. Repeat this process so that the opposite of each of the eight situations is described.

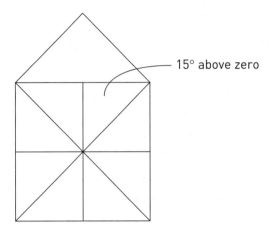

15° above zero

- Refold the paper.

5. Instruct your students to use their fortune tellers to ask a partner (or a classmate if they are working individually) to choose "Left," "Up," "Right," or "Down." Explain that they should do the following:

- Flap the fortune teller while spelling out the word their partner chooses.

- Ask their partner to choose one of the situations that is revealed and ask him to identify the opposite of the situation.

- Lift the flap to see if he is correct.

6. Students should switch roles and continue.

CLOSURE

Ask your students to share some situations and opposites. Ask them to explain the meaning of zero in each of the situations.

The Number System: 6.NS.6

"Apply and extend previous understandings of numbers to the system of rational numbers."

6. "Understand a rational number as a point on the number line. Extend number line diagrams and coordinate axes familiar from previous grades to represent points on the line and in the plane with negative number coordinates.

a. "Recognize opposite signs of numbers as indicating locations on opposite sides of 0 on the number line; recognize that the opposite of the opposite of a number is the number itself, e.g., $-(-3) = 3$, and that 0 is its own opposite.

b. "Understand signs of numbers in ordered pairs as indicating locations in quadrants of the coordinate plane; recognize that when two ordered pairs differ only by signs, the location of the points are related by reflections across one or both axes.

c. "Find and position integers and other rational numbers on a horizontal or vertical number line diagram; find and position pairs of integers and other rational numbers on a coordinate plane."

BACKGROUND

Every rational number can be graphed on a number line. If the number line is horizontal, positive numbers are located to the right of zero and negative numbers are located to the left of zero. If the number line is vertical, positive numbers are above zero and negative numbers are below zero. Numbers that have opposite signs are graphed the same distance from zero, but in opposite directions.

A coordinate plane is formed when a horizontal number line and a vertical number line intersect at a right angle. The point where the two number lines intersect is the origin. It is represented by (0, 0). Every ordered pair can be graphed in a coordinate plane.

ACTIVITY 1: GRAPHING ON A NUMBER LINE

Students will find, record, and graph on a number line the record high temperatures and record low temperatures of two states in the United States. This activity may be done in class or at home.

Rulers; blank paper; markers or colored pencils; current almanacs and other reference books; computers with Internet access.

PREPARATION

Make a "Student Sign-Up Sheet for States" listing each state to serve as a record of the states students will research.

PROCEDURE

1. Explain that each student will sign up to research the record high temperature and record low temperature of two states of the United States. (If you have fewer than 25 students in your class, perhaps some students can research an additional state. If you have more than 25 students, some students may research the same states.)

2. Assign, or allow students to select, the states they will research. Having them write their names next to their states on the "Student Sign-Up Sheet for States" is a simple way to record which students are doing which states.

3. Suggest that students use reference materials or the Internet to find their information. They may search the Internet with terms such as "highest and lowest temperature in U.S. states." They may search specific states as well. They should record the dates of the highest and lowest temperatures, as well as the source of their data.

4. Instruct your students to record all temperatures in degrees Fahrenheit. If they find a temperature in degrees Celsius, they should use the formula $F = 1.8C + 32$ to convert the temperature to the Fahrenheit scale.

5. Students should construct either a horizontal or vertical number line by finding the range of values, choosing a scale for each interval, and then graphing each temperature on the number line. They should label the corresponding values for the highest and lowest temperatures of their state and the name of the state. Students should also record the date of the temperatures beneath the temperature on their number line.

CLOSURE

Display students' number lines in your classroom. Ask your students to identify both the highest and lowest temperatures in the United States. You might also ask them to look for patterns. For example, how does elevation affect temperature? Which states, in general, have the highest "high" temperatures, and which have the lowest "low" temperatures. What are some of the biggest ranges of low and high temperatures?

ACTIVITY 2: BONK THE MOLE

Students will complete this activity on a Web site. They will find and bonk a mole by identifying the mole's coordinates on the coordinate plane. Most students will enjoy bonking the mole as they are learning about the coordinate plane.

MATERIALS

Computers with Internet access.

PROCEDURE

1. Instruct your students to go to the Web site http://funbasedlearning.com/algebra /graphing/default.htm.

2. Explain that they will play a game where they must "bonk" a mole before he eats vegetables. To bonk him, they must identify his coordinates.

3. Review the coordinate plane by instructing your students to click on "Easy Version of Graph Mole." Instruct them to click on "Introduction," read the introduction, then play the game.

4. Once your students have mastered the easy version of the game, encourage them to try the medium and harder versions.

CLOSURE

Ask your students questions such as the following: In what quadrant are both coordinates positive? (Quadrant I) In what quadrant are both coordinates negative? (Quadrant III) In what quadrant is x positive and y negative? (Quadrant IV) In what quadrant is x negative and y positive? (Quadrant II) If two ordered pairs differ by the sign of the x-coordinates, the points are related by a reflection across an axis. Which axis are the points reflected across? (the y-axis)

The Number System: 6.NS.7

"Apply and extend previous understandings of numbers to the system of rational numbers."

7. "Understand ordering and absolute value of rational numbers.

   a. "Interpret statements of inequality as statements about the relative position of two numbers on a number line diagram.

   b. "Write, interpret, and explain statements of order for rational numbers in real-world contexts.

   c. "Understand the absolute value of a rational number as its distance from 0 on the number line; interpret absolute value as magnitude for a positive or negative quantity in a real-world situation.

   d. "Distinguish comparisons of absolute value from statements about order."

BACKGROUND

Every real number can be represented as a point on a number line. A point labeled 0 is called the origin of the number line. Points to the right of 0 represent positive numbers; points to the left of 0 represent negative numbers. Integers, fractions, and decimals can be graphed on a number line. The absolute value of a number is the distance a number is from 0 on the number line. Because a distance is always a positive number, the absolute value of any number, except 0, is always positive.

A number line is divided into equally spaced intervals. For most number lines, each interval represents an integer. To graph a number on a number line, mark the point on the number line that is paired with the number.

 ACTIVITY: AN OLD-FASHIONED NUMBER LINE

Working in groups of three or four, students will create number lines and display points that relate the ordering and absolute value of rational numbers. After completing their number lines, they will share their work with the class.

MATERIALS

Poster paper; rulers; different-colored markers.

1. Explain to your students that the ordering of numbers and absolute values can be represented on a number line. "Seeing" the position of numbers and values on a number line can help in understanding numerical relationships.

2. Offer this example to your students: $-5 < 8$. This can be shown on the number line by graphing -5 and then graphing 8. This can be stated as -5 is to the left of 8 or 8 is to the right of -5.

3. Write the following statements on the board:

    - $-2 > -3$
    - $3 < 4$
    - $-\frac{1}{2} < 0$
    - $-8 > -8\frac{1}{2}$
    - $5°F > 0°F$
    - 25 mph > 15 mph
    - 10 feet below sea level < 2 feet below sea level
    - $|-12| > 10$
    - A \$20 debt < a \$14 credit
    - 3 seconds before liftoff < 3 seconds after liftoff

4. Explain that students are to draw a number line on poster paper. They are then to interpret the statements above and graph the numbers on their number line. They should also create two inequalities of their own and graph these as well. On a separate sheet of paper, they are to include brief descriptions as to why they placed these points as they did, mentioning the location of the numbers on the number line and how the location relates to the inequality, using the example you provided earlier as a guide.

5. Offer these suggestions for creating number lines:

    - Create a number line with a range of -25 to 25. The origin, 0, should be located in the middle of the number line.

    - Use rulers to draw an accurate number line. Make sure the units on the number line are equidistant. Students may include a tick mark for every unit, but instead may prefer to label numbers in multiples of 5.

    - Print neatly. Use markers with bold colors. Make the number line appealing and clear.

CLOSURE

Have students share their posters with the class. Discuss the posters and review the statements the groups generated. Display the posters.

The Number System: 6.NS.8

"Apply and extend previous understandings of numbers to the system of rational numbers."

> 8. "Solve real-world and mathematical problems by graphing points in all four quadrants of the coordinate plane. Include use of coordinates and absolute value to find distances between points with the same first coordinate or the same second coordinate."

BACKGROUND

The coordinate plane is formed by the intersection of a horizontal number line, the *x*-axis, and a vertical number line, the *y*-axis. The two lines intersect at their zero points, which is called the origin and is written as the ordered pair (0, 0). Every ordered pair consists of an *x*-coordinate and *y*-coordinate and can be graphed in the coordinate plane. The *x*-coordinate tells how many places horizontally from zero, or the origin, a point is located. The *y*-coordinate tells how many places vertically from zero a point is located. The coordinate plane is divided into four quadrants, or sections, that are formed by the intersection of the *x*-axis and *y*-axis. The quadrants are labeled counterclockwise beginning with Quadrant I in the upper right and ending with Quadrant IV in the lower right. The coordinate plane is used to locate points and calculate distances between them.

 ## ACTIVITY: THE MAZE GAME

Students will work at a Web site where they will use an interactive game to move a robot through a mine field to its target. Students must correctly input ordered pairs to direct the robot's movements.

MATERIALS

Graph paper; computers with Internet access.

PROCEDURE

1. Review concepts of the coordinate plane with your students. They should be familiar with locating ordered pairs before beginning this activity.

2. Instruct your students to go to the Web site at http://www.shodor.org/interactivate /activities/MazeGame/.

3. Explain that they will play The Maze Game, the goal of which is to move a robot from its starting location by entering the *x*-coordinate and *y*-coordinate of the robot's next step to the target without hitting any mines. Note that students can only move the robot vertically or horizontally. As students work through this activity, they should record the robot's movements on graph paper. For each move horizontally and vertically, students should record the distance on their graph paper. For example, if the robot starts at $(-5, 3)$ and a student moves it to $(-5, -5)$, the student would graph these points on her graph paper. Then she would record a distance of 8 units (found by subtracting the *y*-coordinates and the absolute value) along that line to show that the robot moved 8 units. She would continue to do this for each move until the robot reached its target.

4. Explain that when students have passed one mine field, they may continue this procedure advancing through the next levels of mines. (They can select 5, 10, 15, 20, 25, or 30 mines to be placed throughout the coordinate plane.)

5. Allow time for students to complete several levels of the game.

CLOSURE

Discuss this activity as a class. Select a few students to show their graphs to the class and explain their strategies for winning the game. Also, discuss how students found the distances the robot moved. Finally, ask your students how graphing on the coordinate plane can be used to solve real-world problems.

Expressions and Equations: 6.EE.1

"Apply and extend previous understandings of arithmetic to algebraic expressions."

1. "Write and evaluate numerical expressions involving whole-number exponents."

BACKGROUND

An exponent shows the number of times a base is used as a factor. Following are some examples:

- 3^2 is read "3 squared" or "3 to the second power." $3^2 = 3 \times 3 = 9$

- 4^3 is read "4 cubed" or "4 to the third power." $4^3 = 4 \times 4 \times 4 = 64$

- 5^1 is read "5 to the first power" and is equal to 5. Any base raised to the first power is equal to the base.

- 2^0 is read "2 to the zero power" and is equal to 1. Any base (except 0) raised to the zero power is equal to 1.

If negative numbers are used, students may have trouble recognizing whether the base is positive or negative. Following are some examples:

- $(-3)^4$ is read "−3 to the fourth power." $(-3)^4 = -3 \times (-3) \times (-3) \times (-3) = 81$

- -3^4 is read "the opposite of 3^4." $-3^4 = -(3 \times 3 \times 3 \times 3) = -81$

 ## ACTIVITY: FIND WHICH DOES NOT BELONG

Working individually, then in pairs or groups of three, students will decide which expression on the "Expressions Grid" differs from the other two expressions in the same row.

MATERIALS

Reproducible, "Expressions Grid."

PROCEDURE

1. Discuss the meanings of bases and exponents, using the information in the background as a guide. If necessary, provide additional examples.

2. Explain that expressions with exponents may be described numerically or verbally, but that each expression has only one value. For example, "three squared," "three to the second power," and 3^2 all equal 9.

3. Distribute copies of the reproducible, one to each student. Explain that two of the three expressions in each row have the same value. Working alone, students must choose the expression in each row that is different in value from the other two. They are to provide an explanation for their choice.

4. After students have completed the reproducible, instruct them to work with their partner or group and discuss their answers, correcting any wrong answers.

CLOSURE

Provide the answers to the reproducible and discuss any expressions that students found difficult to understand.

ANSWERS

The expression that does not belong is shown, followed by the reason. Row 1: 6; the other expressions equal 8. Row 2: -7^2 equals -49; the other expressions equal 49. Row 3: 9; the other expressions equal 27. Row 4: 6; the other expressions equal 1. Row 5: -1; the other expressions equal 1. Row 6: 8^2 equals 64; the other expressions equal 16. Row 7: -25; the other expressions equal 25. Row 8: 4^0 equals 1; the other expressions equal 4. Row 9: -100; the other expressions equal 100. Row 10: 100; the other expressions equal -100. Row 11: 5^2 equals 25; the other expressions equal 32. Row 12: 2^3 equals 8; the other expressions equal 1.

EXPRESSIONS GRID

Row #			
1	2^3	2 to the third power	6
2	(−7) squared	$−7^2$	49
3	3^3	9	27
4	6	1^6	6^0
5	$(−1)^2$	$(−1)^0$	−1
6	8^2	4^2	16
7	The opposite of −5, squared	25	−25
8	4^1	4^0	4
9	10^2	−100	10 squared
10	$−10^2$	−100	100
11	5^2	2^5	32
12	8^0	2^3	1

Expressions and Equations: 6.EE.2

"Apply and extend previous understandings of arithmetic to algebraic expressions."

2. "Write, read, and evaluate expressions in which letters stand for numbers.

   a. "Write expressions that record operations with numbers and with letters standing for numbers.

   b. "Identify parts of an expression using mathematical terms (sum, term, product, factor, quotient, coefficient); view one or more parts of an expression as a single entity.

   c. "Evaluate expressions at specific values of their variables. Include expressions that arise from formulas used in real-world problems. Perform arithmetic operations, including those involving whole-number exponents, in the conventional order when there are no parentheses to specify a particular order (Order of Operations)."

BACKGROUND

There are two types of expressions: numerical expressions and algebraic expressions. A numerical expression names a number. Examples include $3 + 4$, $2(3 + 7)$, $18 \div 6$, and 104×5. An algebraic expression is an expression that contains a variable; for example, $2x + 3$, $3(4x - 5)$, and $\frac{1}{3}x$.

To evaluate a numerical expression, students must follow the order of operations. To evaluate an algebraic expression, they must substitute a value for the variable, then use the order of operations.

An equation is a statement that two expressions are equal. Students may solve equations by first evaluating the expression or expressions on one or both sides of the equation.

 ACTIVITY: AND IT EQUALS...

Students will be given a slip of paper that contains an expression or formula and a number or expression. Working individually or in pairs, students will identify the value or expression that matches an expression or formula.

MATERIALS

Reproducible, "Expressions, Equations, and Values."

Make one copy of the reproducible. (You may prefer to enlarge the reproducible before photocopying.) Cut out each box so that you have a total of 21 slips of paper. The original will serve as your answer key. The slips are arranged in order on the reproducible, each providing the answer that correctly matches an expression or formula written on the previous slip, except the match to the last formula which is written on the first slip.

PROCEDURE

1. Mix the slips up, then distribute one slip of paper to each student (or a slip to pairs of students). For a small class, you may give some students two slips. You must distribute all 21 slips.

2. To start, choose a student to read the expression or formula that is written on the right side of his slip. You may find it helpful to write the expression or formula on the board. All students should try to find the value or expression on the left side of their slips that matches the expression or formula that was presented. Because of the way the slips are designed, only one will contain a correct match. The student who has the slip with the correct answer should say "I have . . . " and then provide the answer. If the student is correct, he then reads the expression or formula written on the right side of his slip. If he is incorrect, point out his error. Another student should then provide the correct answer printed on the left side of her slip.

3. Continue the process until the student who read the first formula has the match to the last formula.

CLOSURE

Discuss the activity. Did students find other answers in addition to those on the cards? Explain that an expression may be written in several ways.

EXPRESSIONS, EQUATIONS, AND VALUES

I have 26	Find P ; $s = 1.5$ $P = 4s$	I have 6	Find A ; $s = 1.5$ $A = s^2$	I have 2.25	Three less than a number
I have $n - 3$	The quotient of a number and 4	I have $\frac{n}{4}$	The product of 3 and a number	I have $3n$	The coefficient of $4n$
I have 4	Two factors of $2(n + 3)$	I have 2 and $(n + 3)$	A number that represents $(3 - 12)$	I have -9	The sum of a number and 3
I have $n + 3$	Find A ; $s = 1.5$ $A = 6s^2$	I have 13.5	Find V ; $s = 1.5$ $V = s^3$	I have 3.375	$x = 2$ Find $8x^2$
I have 32	The terms of $(-1 + \frac{3}{4})$	I have -1 and $\frac{3}{4}$	Find the value of $19 - 5 - 2$	I have 12	Six times a number
I have $6n$	A number divided by $\frac{1}{4}$	I have $4n$	Find the value $3^2 - 1$	I have 8	Find the value of $3 + 8 \times 2$
I have 19	Find the value of $3(-6 + 2)$	I have -12	$x = \frac{1}{2}$ Find $-2x^2$	I have $-\frac{1}{2}$	Find C ; $t = 4$ $C = 6.5t$

Expressions and Equations: 6.EE.3

"Apply and extend previous understandings of arithmetic to algebraic expressions."

3. "Apply the properties of operations to generate equivalent expressions."

BACKGROUND

Several properties are true for all rational numbers. Some of the most important include:

- Commutative property of addition: $a + b = b + a$
- Commutative property of multiplication: $ab = ba$
- Associative property of addition: $(a + b) + c = a + (b + c)$
- Associative property of multiplication: $a \times (b \times c) = (a \times b) \times c$
- Distributive property: $a(b + c) = ab + ac$ and $(b + c)a = ba + ca$
- Addition property of zero: $a + 0 = 0 + a = a$
- Sum of opposites property: $a + (-a) = -a + a = 0$
- Multiplication property of one: $a \times 1 = 1 \times a = a$
- Multiplication of reciprocals property: $a \times \dfrac{1}{a} = \dfrac{1}{a} \times a = 1, a \neq 0$

Understanding properties helps students not only to compute accurately but also to clarify their mathematical reasoning and recognize mathematical relationships.

 ## ACTIVITY: PRESENTING PROPERTIES

For this activity, students will work in groups of four or five. They will create a presentation of a property of mathematics, explain the property and include various examples of how the property can be applied to generate equivalent expressions. This activity will require two class periods—the first for students to create and practice their presentations; the second for students to make their presentations to the class.

MATERIALS

Depending on the presentations, computers; digital projector; whiteboard or screen; overhead projector; transparencies; markers; math texts; math reference books. Optional: computers with Internet access.

1. Explain that each group is to create a presentation in which they will explain a mathematical property to the class. (Depending on the size of your class, you may prefer that each group of students presents only one property, for example, the commutative property of addition, or two properties, for example, the commutative properties of addition and multiplication.) Note that students will also include examples that show how the property can be applied to generate equivalent expressions. Offer the following examples of equivalent expressions of the associative property of addition:

$$(3 + 6) + 7 = 3 + (6 + 7)$$

$$9 + 7 = 3 + 13$$

$$16 = 16$$

$$\left(\frac{3}{10} + \frac{1}{10}\right) + \frac{7}{10} = \frac{3}{10} + \left(\frac{1}{10} + \frac{7}{10}\right)$$

$$\frac{4}{10} + \frac{7}{10} = \frac{3}{10} + \frac{8}{10}$$

$$\frac{11}{10} = \frac{11}{10}$$

$$1\frac{1}{10} = 1\frac{1}{10}$$

2. Explain that the presentations may take various forms, for example: a mini-lesson that the members of the group teach to the class; a skit (perhaps a group member assuming the role of a teacher and the other members assuming roles of students); or a demonstration (perhaps in which group members explain their property and provide examples in the form of charts or posters). Encourage students to use materials that will help them to make clear and informative presentations. Caution them that any props they use should be simple, safe, and easily obtainable.

3. Encourage your students to research their properties using their math texts and math reference books. Researching properties on the Internet (using the property's name as a search term) is an option. Emphasize that the information in their presentations must be accurate.

4. During their presentations, students should explain their property and provide examples. Encourage all group members to participate in their group's presentation.

CLOSURE

Have a member of each group briefly summarize the group's property for the class.

Expressions and Equations: 6.EE.4

"Apply and extend previous understandings of arithmetic to algebraic expressions."

4. "Identify when two expressions are equivalent (i.e., when the two expressions name the same number regardless of which value is substituted into them)."

BACKGROUND

Expressions are mathematical symbols that represent a number. An expression can be numerical or algebraic:

- A numerical expression contains only numbers and operations, and/or grouping symbols. A numerical expression names a particular number. Examples: $15 - 8$; 6×5; $4(3 + 2)$; $1\frac{1}{2} \div \frac{3}{4}$

- An algebraic expression includes a variable or variables. It may also include numbers, operations, and/or grouping symbols. Two algebraic expressions are equivalent if they name the same number, no matter what value is substituted for the variable. Examples: $8n$; $4n + 7$; $4(x - y)$; $\frac{y}{18}$

 ACTIVITY: PARTNER QUIZ

Students will first work individually, then work with a partner or in a group of three. They will create quizzes of equivalent expressions that their partner must solve.

PROCEDURE

1. Explain that expressions are mathematical symbols that represent a number, or can represent a number if values are provided for the variables. Also explain the difference between numerical expressions and algebraic expressions, offering the examples that are included in the Background for this activity. Note that for this activity, students will focus on algebraic expressions.

2. Provide examples of equivalent algebraic expressions such as: $x + x + x + x = 4x$; $\frac{1}{2}y = \frac{y}{2}$; and $3(x + 1) = 3x + 3$.

3. Explain to your students that they are to create a quiz that focuses on algebraic expressions. Suggest that they consider a "matching" quiz set up in two columns. (They can create other forms of quizzes, but a simple quiz in which expressions are matched is easy to do, freeing students to concentrate on the math.) In the first column, they may write five algebraic expressions, numbered 1 through 5. In the second column, they should write an equivalent algebraic expression (not in the same order) for each expression in the first column. Instead of numbers, they should designate each expression in the right column by a letter. On the back of their paper, they should write the correct answers.

4. After students have created their quizzes, they are to exchange their quizzes with another student. They then solve each other's quizzes by matching the correct algebraic expressions.

5. After students have completed each other's quizzes, they should return and correct them.

CLOSURE

Ask for volunteers to share what they felt were particularly challenging problems. Write these problems on the board and discuss them as a class. Also correct any problems whose answers are in dispute.

Expressions and Equations: 6.EE.5

"Reason about and solve one-variable equations and inequalities."

5. "Understand solving an equation or inequality as a process of answering a question: which values from a specified set, if any, make the equation or inequality true? Use substitution to determine whether a given number in a specified set makes an equation or inequality true."

BACKGROUND

Equations and inequalities may be true or false, depending on the values that are substituted for variables. If a value makes an equation or inequality true, then the value is a solution to the equation or inequality.

ACTIVITY: THREE IN A ROW

Working in pairs or groups of three, students will rearrange equations and inequalities so that each equation or inequality in the same row matches a specific solution.

MATERIALS

Scissors; glue sticks; reproducible, "Equations and Inequalities."

PREPARATION

Make two copies of the reproducible for each pair or group of students.

PROCEDURE

1. Explain that a solution to an equation or inequality is a value that makes the statement true. Students can determine if a value is a solution to an equation or inequality by substituting the values of the variable into the equation or inequality.

2. Distribute two copies of the reproducible to each pair or group of students.

3. Explain that they are to cut out each of the equations and inequalities on one of the reproducibles. They will have 24 separate cards containing either equations or inequalities. Using the second copy of the reproducible, they are to work together to match three cards with the solution provided in each row. (On the original, the equations and inequalities do not match the solutions in their rows.) Students should place a matching equation or inequality over an original equation or inequality on the reproducible. Once students have agreed they have placed each card in the correct row, they should glue the cards. When they are finished, the solution of each row should be a solution to the equations and inequalities that follow it. Note that there is only one way to arrange the cards.

CLOSURE

Discuss the answers as a class.

ANSWERS

Answers for each row may be in any order: **Row 1:** $-2n = -16$; $9 + n = 17$; $-3n = -24$. **Row 2:** $25 - n \geq 5$; $2n = 40$; $3 + n < 25$. **Row 3:** $36 = n + 10$; $2n \geq 52$; $24 = n - 2$. **Row 4:** $2n = 0$; $15 - n = 15$; $0 \div 10 = n$. **Row 5:** $-16 \div -4 = n$; $5n < 24$; $3n = 12$. **Row 6:** $6 \div n = 3$; $-2n = -4$; $8n = 16$. **Row 7:** $48 \div 4 = n$; $2n = 24$; $12 \div n = 1$. **Row 8:** $7 = 8 + n$; $9 - 10 = n$; $3n \leq -3$.

Row #	Solution	Equations and Inequalities		
1	$n = 8$	$3n \leq -3$	$7 = 8 + n$	$-16 \div -4 = n$
2	$n = 20$	$5n < 24$	$9 - 10 = n$	$48 \div 4 = n$
3	$n = 26$	$12 \div n = 1$	$2n \geq 52$	$-2n = -4$
4	$n = 0$	$2n = 24$	$8n = 16$	$-2n = -16$
5	$n = 4$	$2n = 40$	$3n = 12$	$3 + n < 25$
6	$n = 2$	$-3n = -24$	$25 - n \geq 5$	$36 = n + 10$
7	$n = 12$	$2n = 0$	$9 + n = 17$	$24 = n - 2$
8	$n = -1$	$15 - n = 15$	$0 \div 10 = n$	$6 \div n = 3$

Expressions and Equations: 6.EE.6

"Reason about and solve one-variable equations and inequalities."

> 6. "Use variables to represent numbers and write expressions when solving a real-world or mathematical problem; understand that a variable can represent an unknown number, or, depending on the purpose at hand, any number in a specified set."

BACKGROUND

A variable represents an unknown quantity. It is usually expressed as a letter. An expression is a group of mathematical symbols, for example, numbers, variables, and operations that represent a number, or can represent a number if values are assigned to the variables. Variables and expressions are essential for solving mathematical problems.

ACTIVITY: A SLICE OF LIFE WITH VARIABLES AND EXPRESSIONS

Students are to write a story in which a problem can be modeled mathematically. They will model the problem using an expression that contains at least one variable.

MATERIALS

Optional: computers and printers.

PROCEDURE

1. Explain to your students that a "slice of life" is a short piece of writing that depicts a relatively minor but nonetheless important aspect of life. For this activity, students are to write a slice-of-life piece in which math plays a part. They will use variables and expressions to model the math in their writing.

2. Explain that an algebraic expression contains a variable. Expressions can be used to model and solve problems. For example, Maria is 5 years older than her brother Paulo, who is 8 years old. Maria's age, represented by the variable M, can be expressed as $M = 8 + 5$. Here is another example. Alicia is 10 years old. She is four years older than her brother. Alicia's age may be expressed as 10 or $b + 4$, where b represents the age of Alicia's brother.

3. Instruct your students to write a slice-of-life story. Their story may be entirely fictional or it may be based on a real-life event. Their story should include a problem or situation that can be modeled and solved by using a mathematical expression. Brainstorm with your students to generate some possible ideas that can help them get started in developing ideas for their stories.

   - A trip to the mall to buy new clothes and eat at the food court

   - A visit to an amusement park

   - A part-time job

   - The school spring dance

   - A sporting event

   - A favorite hobby or activity

4. Offer this excerpt of a possible story to your students. "I babysit each week for my neighbor's little girl, Rachel. Last week I earned $50 for working five hours. I read to Rachel, we colored, and we watched a little bit of TV." Ask your students how they might model the writer's earnings. One possibility is $n =$ the hourly wage, therefore $5n$ is an expression that represents the amount of money earned. $5n = \$50$ is an equation that models this situation. The solution is $n = \$10$.

5. Encourage your students to use correct grammar and mechanics in their stories, as well as be mathematically accurate.

6. After students have completed their writing, they should select an event of the story they will model mathematically using expressions and variables. They should write the expression, and the meaning of each variable, at the end of their story.

CLOSURE

Ask volunteers to read their stories to the class and explain how math describes a situation in the story. Discuss how math can be used to model events in life. Display the stories.

Expressions and Equations: 6.EE.7

"Reason about and solve one-variable equations and inequalities."

> 7. "Solve real-world and mathematical problems by writing and solving equations of the form $x + p = q$ and $px = q$ for cases in which p, q, and x are all nonnegative rational numbers."

BACKGROUND

Mathematical equations are commonly used to solve real-world problems. When students use equations, they must understand that variables stand for letters, and they must understand the steps necessary for solving one-step equations.

To solve equations in the form of $x + p = q$, students should subtract the value of p from both sides of the equation.

To solve equations in the form of $px = q$, students should divide both sides of the equation by the value of p, provided that the value of p is not equal to zero.

ACTIVITY: EQUATIONS, EQUATIONS, EQUATIONS

Students will write problems and use equations to solve problems. You will present individual student's problems to the class to solve.

MATERIALS

One transparency per student; markers; tissues or erasers.

PROCEDURE

1. Explain to your students that you will provide them with various real-life situations that they will use as the foundation for writing problems that can be solved using the equations $x + p = q$ and $px = q$. x, p, and q are nonnegative, rational numbers. If necessary, review the steps for solving these types of equations.

2. List the following examples of situations and encourage your students to offer more:

   * The Little League bake sale

   * A party

   * Going to a movie

- Shoveling snow for neighbors for a fee

- Saving money to buy a new laptop

- Taking public transportation

- Going shopping

3. Explain that each of these broad situations presents possibilities for writing word problems that can be solved using the equations stated above. For example, a store offers a $5 refund on your next purchase if you buy three bottles of shampoo. Sheila has purchased two bottles of shampoo. How many more bottles does she need to qualify for the refund? This problem can be solved using the equation $x + p = q$ where x equals the number of bottles left to purchase, p equals the bottles of shampoo Sheila purchased, and q equals the amount of bottles needed to qualify for the refund. $x + 2 = 3$. $x = 1$. Here is another example. A supermarket offers a discount for each shopping bag customers bring to the store to bag their purchases. Mike brought four bags to the store and saved 20 cents. What was the discount? This problem can be solved using the equation of the form $px = q$ where p stands for the number of bags, x stands for the discount per bag, and q stands for the discount. This can be written as $4x = \$0.20$. $x = \$0.05$. You might want to point out to your students that at first glance, these equations might be considered easy; however, they serve as a means for helping students to recognize real-life situations that may be expressed mathematically.

4. Explain that students are to choose one or two real-life situations, either those you listed or their own, and write two problems, one that can be solved using the equation $x + p = q$ and the other using the equation $px = q$. Students should write their problems near the top of the transparency. At the bottom of the transparency, they should write the equations that can be used to solve the problems, the solutions to the problems, and their names. Remind them to write neatly.

5. After students have completed their problems, collect their transparencies. Present some of your students' problems to the class, keeping the equations and students' names on the bottom of the transparencies covered. The class should solve the problem. While keeping the name on the transparency covered, compare the answer on the transparency to the answer the class believes is correct.

CLOSURE

Discuss how equations can be used to solve problems. Ask students to describe the steps they used for solving problems with the two equations they used in this activity.

Expressions and Equations: 6.EE.8

"Reason about and solve one-variable equations and inequalities."

8. "Write an inequality of the form $x > c$ or $x < c$ to represent a constraint or condition in a real-world or mathematical problem. Recognize that inequalities of the form $x > c$ or $x < c$ have infinitely many solutions; represent solutions of such inequalities on number line diagrams."

BACKGROUND

An inequality is a mathematical expression that shows two quantities are not equal. The following symbols are used in inequalities:

- Greater than: $>$
- Less than: $<$
- Greater than or equal to: $\geq$
- Less than or equal to: $\leq$
- Not equal to: $\neq$

(Note: This Standard only addresses "greater than" or "less than" inequalities; however, students should also understand the symbols for "greater than or equal to," "less than or equal to," and "not equal to.")

An inequality can have infinitely many solutions. The solution to an inequality is the value or values that make the inequality true. For example, the solutions to the inequality $x > 5$ are all real numbers that are larger than 5.

The solutions to inequalities can be represented on a number line. The following guidelines summarize the steps for graphing inequalities that contain the symbols $>$ and $<$.

- Rewrite the inequality so that the variable is isolated on the left in the number sentence (if necessary).
- Draw a number line and label it with the appropriate numbers.
- For inequalities that are greater than or less than a number, the number is not included in the solution set. Draw an open circle on the point that corresponds to the number on the number line.
- Draw an arrow along the number line, pointing to the numbers that are included in the solution set.

Example: $x > 5$

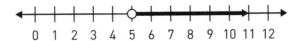

5 is not included in the solution set and is marked as an open circle. The arrow points to the right because numbers to the right of 5 satisfy the inequality.

Example: $x < 2$

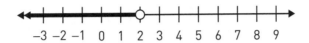

2 is not included in the solution set and is marked as an open circle. The arrow points to the left because numbers to the left of 2 satisfy the inequality.

(Note: The procedure for graphing inequalities that contain $\geq$ and $\leq$ is the same as the examples above, except that the circle is closed instead of open.)

 ACTIVITY: FIND YOUR MATCH

Students will be given a card that describes a situation that can be modeled by an inequality or a card that contains the graph of an inequality. Students who have the description card must write an inequality using symbols and find the graph that matches the card they have. At the same time, students who have the graph card must write the inequality that is graphed and find the description of the situation modeled by the graph.

MATERIALS

Index cards; 2-page reproducible, "Inequality and Graph Cards."

PREPARATION

Make one copy of the reproducible and cut out the description cards and the graph cards. Use the original as an answer key. Each graph is to the right of the situation it models.

PROCEDURE

1. Review the concepts of inequalities with your students, including how to graph inequalities on a number line.

2. Explain that half of the students in class will receive a card that contains a description of an inequality. (If you have an odd number of students, one student may be given two cards.) The other half of the class will receive graph cards that correspond to the inequalities.

3. Randomly distribute the description cards and the graph cards, one per student. Instruct them to write an inequality to describe the situation or graph, depending on the type of card they received. Instruct them to use x to represent the variable.

4. Instruct students to walk around the room and sit with the person who has a matching card.

5. Once all students have found their match, select a few students to share their descriptions, the inequalities they wrote, and their graphs.

6. You may repeat this activity by shuffling the cards and passing them out to students again.

CLOSURE

Provide each of your students with an index card. On the front of the card, they are to write a situation that can be expressed as an inequality. On the back, they are to graph the inequality they wrote. Collect the cards as your students leave. You may wish to use these as a review or follow-up activity.

INEQUALITY AND GRAPH CARDS

Descriptions Cards	Graph Cards
To buy a ticket for the county fair, you must have more than $5.	3 4 5 6 7 8
To enter the kids' pool, you must be less than 5 feet tall.	3 4 5 6 7 8
You must order more than 30 cupcakes for your birthday party.	28 29 30 31 32 33
Luckily, your math test has fewer than 30 questions.	28 29 30 31 32 33
Your walk to school is more than 0.5 mile.	−2 −1 0 1 2 3
Gym class is less than a half hour.	−2 −1 0 1 2 3
To receive an A in Social Studies, you need a grade greater than 85% on your test.	82 83 84 85 86 87
The weather person said the temperature would be lower than 85°F today.	82 83 84 85 86 87
Your older sister gets paid more than $45 for tutoring.	42 43 44 45 46 47
You must be more than 4.5 feet tall to ride a roller coaster.	1 2 3 4 5 6

(continued)

Descriptions Cards	Graph Cards
The admission price of the matinee is less than $4.50.	←—┼——┼——○——┼——┼——┼——→ 2 3 4 5 6 7
The blizzard brought more than 3 feet of snow!	←—┼——┼——○━━┼━━┼━━┼━→ 1 2 3 4 5 6
There are fewer than 3 days of school left!	←━━┼━━┼━━○——┼——┼——┼—→ 1 2 3 4 5 6
I went to bed after 10:00 last night.	←—┼——┼——○━━┼━━┼━━┼━→ 8 9 10 11 12 13
My birthday is fewer than 10 days away!	←━━┼━━┼━━○——┼——┼——┼—→ 8 9 10 11 12 13

Expressions and Equations: 6.EE.9

"Represent and analyze quantitative relationships between dependent and independent variables."

> 9. "Use variables to represent two quantities in a real-world problem that change in relationship to one another; write an equation to express one quantity, thought of as the dependent variable, in terms of the other quantity, thought of as the independent variable. Analyze the relationship between the dependent and independent variables using graphs and tables, and relate these to the equation."

BACKGROUND

A variable is defined as a quantity that can change. It can be an event, object, measurement, or any other item that can be measured. There are two types of variables: independent and dependent. An independent variable is the variable that determines the value of the other variable. It is the variable that stands alone and is not affected by changes in the other variable. For example, in the formula $d = 45t$, t represents time, which is an independent variable. A dependent variable is the variable that depends upon the value of the independent variable. In the formula $d = 45t$, d represents distance and is the dependent variable because distance depends on the length of time traveled. In an experiment, the dependent variable is usually the variable that is being tested or measured. The relationship between these variables can be expressed in tables, graphs, and equations.

ACTIVITY 1: EXAMINING RELATIONSHIPS

Working in groups of three or four, students will be given two variables that they must analyze. They will show the relationship in a table, graph, and equation.

MATERIALS

Graph paper; rulers.

PROCEDURE

1. Assign each group one of the following situations:

   • Joey can run 1 mile in 8 minutes.

   • Sam works at a pizza shop and earns $9 an hour.

- Kylie's family drives to Florida at an average speed of 50 miles per hour.

- Rebecca swims 3 laps in 2 minutes.

- Tim's cell phone bill costs $0.49 per minute for long distance.

2. Instruct each group to use the information presented to them to identify the independent and dependent variables. They will then create a table to show the relationship between the variables by selecting ten values for the independent variable. Using the information in their table, they will construct a graph of their data, making sure to label the independent and dependent variables. Lastly, students will write an equation showing the relationship between the variables.

CLOSURE

Have students present their work to the class. Students should explain how they determined which variable was independent and which was dependent. They should also discuss how they created the table, graph, and equation and how each represents the relationship between the two variables. Discuss each situation, describing how one variable changes in relation to another.

ACTIVITY 2: COMPARING HEART RATES

Students will compare their resting heart rate to their heart rate after 1 minute of exercise. They will organize their results in a table and graph, and write an equation.

MATERIALS

Graph paper; stopwatch.

PROCEDURE

1. Explain to your students that a person's heart rate is the number of times his heart beats per minute. Explain that in this activity, students will find their resting heart rate and compare it to their heart rate after a minute of exercise. Explain that a resting heart rate is the number of times a heart beats in one minute when a person is at rest.

2. Instruct your students to calculate their resting heart rate. They can find their pulse in one of two places: the wrist or neck. Instruct them to place their second and third fingers either on the palm side of the other wrist right below the thumb or on their lower neck just below the windpipe. Tell your students to apply a small amount of pressure until they feel their pulse. Once everyone has found their pulse, tell students to begin counting while you keep track of the time for 10 seconds. After 10 seconds, have

your students write the number they counted on their paper. Because heart rate is defined as the number of heartbeats per minute, have your students multiply that number by 6 to calculate their resting heart rate for 1 minute.

3. Explain that your students are going to conduct an experiment to determine how their resting heart rate compares to their heart rate after 1 minute of exercise. Watch the clock and instruct your students to do jumping jacks. They may jump as fast or as slowly as they want. The purpose is to simply elevate their heart rate to identify a difference between their resting heart rate and their heart rate after exercise. (Caution: Check with your school's nurse before doing any type of exercise with the class. Some students may have a medical condition that prohibits them from exercising. If so, these students can use a classmate's data. You might also allow students to modify jumping jacks to marching in place, if necessary.) After 1 minute, tell your students to stop and immediately find their pulse. Tell them to begin counting while you keep time for another 10 seconds. At the end of 10 seconds, have your students write the number they counted on their paper. Have them multiply that number by 6 to calculate their heart rate per minute after exercise.

4. Using their two heart rates as their data, students should create a three-column table that displays the time (ranging from 1 to 10 minutes), the number of their resting heartbeats, and the number of their heartbeats after exercising. To do this, they must first identify the variables in this situation, which are heartbeats and time.

5. After students have completed their tables, instruct them to graph their data. Depending on the abilities of your students, you may find it necessary to help them select an appropriate scale. Note that each point on the table corresponds to an ordered pair on the graph.

6. Finally, instruct your students to write an equation based on the table and graph for their resting heart rate and heart rate after 1 minute of exercise.

CLOSURE

Select a few students to present their work to the class. Discuss how students determined the independent and dependent variables and how they used their heart rates to create tables, graphs, and equations. Discuss how the relationship between the two variables is represented in the table, graph, and equation. Also discuss what factors may influence heart rates and how this experiment could be applied to real-world situations. (A good example is the individual who exercises and wants to monitor his heart rate.)

Geometry: 6.G.1

"Solve real-world and mathematical problems involving area, surface area, and volume."

> 1. "Find the area of right triangles, other triangles, special quadrilaterals, and polygons by composing into rectangles or decomposing into triangles and other shapes; apply these techniques in the context of solving real-world and mathematical problems."

BACKGROUND

The areas of many two-dimensional figures can be derived from other area formulas. The area of a right triangle, for example, can be found by dividing a rectangle into two congruent parts. Similarly, the area of a trapezoid can be found by dividing a parallelogram into two congruent parts. Realizing how formulas can be derived often makes it easier for students to understand and remember the formulas.

Irregular figures are so varied that the best way to find their areas is to divide them into other geometric figures whose areas can easily be found.

ACTIVITY 1: IT'S HALF

Students will discover the formula for finding the area of a right triangle by using the formula for finding the area of a rectangle.

MATERIALS

Rulers; scissors; rectangular sheets of unlined paper.

PROCEDURE

1. Explain that some area formulas can be found if other area formulas are known. For example, finding the area of a right triangle can be found by using the formula for finding the area of a rectangle.

2. Instruct your students to label their rectangular paper with an "L" for length and a "W" for width.

3. Ask for a volunteer to state the formula for finding the area of a rectangle. ($A = l \times w$)

4. Instruct your students to draw a diagonal line from one vertex of their paper to the opposite vertex. They should cut the paper along the line they drew.

5. Ask what figures are formed. (Two congruent right triangles)

6. Ask how the area of each triangle compares with the area of the rectangle. (Each area is half of the area of the rectangle.)

CLOSURE

Ask your students to write the formula for finding the area of a right triangle. $(A = \frac{1}{2} \times l \times w)$ Review how they derived this formula.

ACTIVITY 2: IT'S AWE-SUM

Students will use virtual interactive geoboards to create irregular polygons. They will find the areas by decomposing the figures into triangles and other shapes.

MATERIALS

Digital projector; computers with Internet access; graph paper.

PROCEDURE

1. Instruct your students to go to the Web site http://nlvm.usu.edu/en/nav/vlibrary.html. They should go to Geometry and click on grades 6–8. Next, they should scroll down and click on Geoboard.

2. Explain to your students that they will use a virtual geoboard to draw irregular polygons. They will then find the area of the figure by dividing it into smaller shapes for which they know area formulas. Note that the sum of the areas of these smaller figures is equal to the area of the irregular polygon.

3. Demonstrate the use of the virtual geoboards.

   • Start by clicking on "Clear" to clear the board.

   • Click on "Bands" and drag a band to the virtual geoboard.

   • Click on the band and drag it to a peg, repeating this process to create an irregular polygon.

4. Instruct your students to make an irregular figure whose sides are line segments on their virtual geoboards. After students have made their figure, ask them to draw it on graph paper. Explain that each line on the graph paper is 1 unit long and the area

of each box is 1 square unit. They can divide the figure they made into triangles, or other figures for which they know area formulas. By adding the areas of these figures, they will be able to find the area of the irregular figure they created. An example is shown below.

 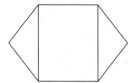 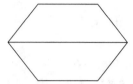

The hexagon can be divided into two triangles and a rectangle or two trapezoids.

5. After they have found the area of their figure on their graph paper, students should go back to their virtual geoboards, click on their figure and then click on "Measure" to find the area of the figure. They can now check that the area they found for their irregular figure is correct. If their answer is incorrect, students should check their sketch and check their work.

6. Allow students time to create additional figures and check their answers.

CLOSURE

Ask your students how this procedure of decomposing irregular figures may apply to real-world problems. (Answers may vary. Finding the area of an L-shaped room is one example.)

Geometry: 6.G.2

"Solve real-world and mathematical problems involving area, surface area, and volume."

2. "Find the volume of a right rectangular prism with fractional edge lengths by packing it with unit cubes of the appropriate unit fraction edge lengths, and show that the volume is the same as would be found by multiplying the edge lengths of the prism. Apply the formulas $V = lwh$ and $V = Bh$ to find volumes of right rectangular prisms with fractional edge lengths in the context of solving real-world and mathematical problems."

BACKGROUND

Volume is the number of cubic units needed to fill a container. The formula for finding the volume of a right rectangular prism (a typical box) is $V = lwh$, where l stands for the length of the rectangular base, w stands for the width of the base, and h stands for the height of the prism. Another formula that gives the same result is $V = Bh$, where B stands for the area of the rectangular base and h stands for the height of the rectangular prism.

 ### ACTIVITY: IT'S VOLUMINOUS

Working in pairs or groups of three, students will "fill" a rectangular prism with cubes by making a sketch, then find the volume of the prism by using formulas. They will also find the volume of other rectangular prisms.

MATERIALS

Rulers; unlined paper; reproducible, "Volumes of Rectangular Prisms."

PROCEDURE

1. To start this activity, explain to your students that they will find the volume of a rectangular prism by sketching cubes that are needed to fill the prism.

2. Distribute a copy of the reproducible to each student. Explain that the figure at the top of the sheet is a rectangular prism, for example, a cardboard box. Students are to imagine filling the box with cubes that measure $\frac{1}{2} \times \frac{1}{2} \times \frac{1}{2}$. To find the volume of the prism they should follow the guidelines under Step 1 on the reproducible.

3. After students have found the volume of the prism in the diagram, they are to complete the rest of the steps on the reproducible. If necessary, review the meaning of the variables in the volume formulas.

Correct and discuss the answers students found.

ANSWERS

1. 55 cubes in an 11 by 5 arrangement of cubes; 7 layers; 385 cubes; $\frac{1}{8}$ cubic inch;

   $385 \times \frac{1}{8} = 48\frac{1}{8}$ cubic inches

2. $5\frac{1}{2} \times 2\frac{1}{2} \times 3\frac{1}{2} = 48\frac{1}{8}$ cubic inches

3. $13\frac{3}{4} \times 3\frac{1}{2} = 48\frac{1}{8}$ cubic inches

4. All of the answers are the same; explanations may vary.

5. $15\frac{5}{8}$ cubic inches; $166\frac{7}{8}$ cubic inches; $131\frac{1}{4}$ cubic inches

VOLUMES OF RECTANGULAR PRISMS

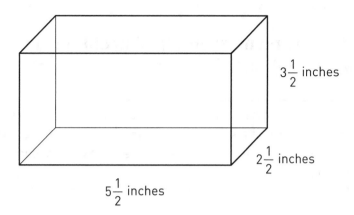

$3\frac{1}{2}$ inches

$2\frac{1}{2}$ inches

$5\frac{1}{2}$ inches

1. Find the volume of the $5\frac{1}{2}$-inch by $2\frac{1}{2}$-inch by $3\frac{1}{2}$-inch rectangular prism by filling it with $\frac{1}{2}$-inch cubes. Follow the guidelines below.

   • Sketch the number of cubes that can be placed in the base of the rectangular prism. This is the number of cubes in one layer. The base measures $5\frac{1}{2}$ inches by $2\frac{1}{2}$ inches. How many cubes, in one layer, would fit in the base?

   • How many layers of cubes are needed to fill a rectangular prism that is $3\frac{1}{2}$ inches high?

   • How many $\frac{1}{2}$-inch cubes are needed to fill the box?

   • What is the volume of each $\frac{1}{2}$-inch cube?

   • Find the volume of the prism by multiplying the total number of cubes by the volume of each cube. What is the volume of the prism?

2. Find the volume of the rectangular prism using the formula $V = lwh$.

3. Find the volume of the rectangular prism using the formula $V = Bh$, where B is the area of the base.

4. How do your answers to questions 1, 2, and 3 compare? Explain your reasoning.

5. Use a formula to find the volume of the following:

   • A cube whose side measures $2\frac{1}{2}$ inches.

   • A cereal box with the following dimensions: $7\frac{1}{2}$ inches by 2 inches by $11\frac{1}{8}$ inches.

   • A container with the following dimensions: $3\frac{3}{4}$ inches by 5 inches by 7 inches.

Geometry: 6.G.3

"Solve real-world and mathematical problems involving area, surface area, and volume."

3. "Draw polygons in the coordinate plane given coordinates for the vertices; use coordinates to find the length of a side joining points with the same first coordinate or the same second coordinate. Apply these techniques in the context of solving real-world and mathematical problems."

BACKGROUND

A polygon is a closed figure whose sides are line segments. A vertex of a polygon is the point where two line segments meet. To draw a polygon in the coordinate plane, students must graph the coordinates of the vertices and draw line segments to connect the vertices.

ACTIVITY: INITIALS

Students will draw their initials represented by polygons in the coordinate plane. They will label the vertices on their graphs and find the lengths of the sides of the polygons used in their initials. After listing their vertices on a separate sheet of paper, students will hand in their lists, which will be randomly redistributed among other students. Students then graph each other's initials in the coordinate plane and confirm the lengths of the sides of the polygons by subtracting the coordinates of the vertices.

MATERIALS

Graph paper; reproducible, "Guidelines for 'Initial' Graphing in the Coordinate Plane."

PROCEDURE

1. Explain that your students will use graph paper to draw their initials in the form of polygons in the coordinate plane.

2. Distribute copies of the reproducible and review the information for graphing points, as well as the steps for finding the lengths of line segments by subtracting the coordinates of the vertices. Note the sample initials, emphasizing how polygons were used to create the initials.

3. Explain that students must use all four quadrants of the coordinate plane when graphing their initials. Note that they may only use polygons to graph their initials. They are to label the vertices on their graphs, and then find the length of each line segment of the polygons that make up their initials. Emphasize that they should only use polygons composed of rectangles and squares with sides that are parallel to an axis in graphing their initials, because this will allow them to find the lengths of the line segments easily by subtracting the appropriate coordinates. They should avoid using diagonal lines.

4. After finding the lengths of the line segments, they are to list the vertices in order on another sheet of paper. Students should not write their names on this sheet.

5. Review the example provided on the reproducible. Make sure that students understand how to list the coordinates of the vertices.

6. After students have completed these steps, they should hand their papers that list vertices to you. After collecting all of the papers, mix them randomly and redistribute them to your students. Explain that students are to now graph the vertices listed on the paper they received, and connect the points in the order the points are listed to form polygons that will reveal the initials of another student. After completing the graphs, they are to verify the lengths of the line segments of the polygons by subtracting the coordinates of the vertices. These second graphs should be identical to the original graphs.

CLOSURE

Ask students to identify the person whose initials they graphed.

GUIDELINES FOR "INITIAL" GRAPHING IN THE COORDINATE PLANE

To graph a point in the coordinate plane, do the following:

1. Start at the origin.

2. Move along the *x*-axis. If the *x*-coordinate is positive, move right. If it is negative, move left. If the *x*-coordinate is zero, do not move.

3. Stop at the point that represents the *x*-coordinate.

4. From this point, move parallel to the *y*-axis. If the *y*-coordinate is positive, move up. If it is negative, move down. If the *y*-coordinate is zero, do not move.

5. Mark the point.

To graph a polygon, connect the vertices in order.

Graph: (0, 3), (−3, 3), (−3, 2), (−2, 2), (−2, −2), (−1, −2), (−1, 2), (0, 2), (0, 3) Stop

Graph: (1, 3), (5, 3), (5, −2), (4, −2), (4, 0), (2, 0), (2, −2), (1, −2), (1, 3) Stop

Graph: (2, 1), (4, 1), (4, 2), (2, 2), (2, 1) Stop

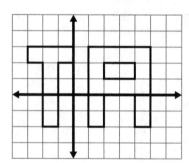

To find the length of a vertical line segment, subtract the *y*-coordinates. (If the value is negative, find the absolute value.)

To find the length of a horizontal line segment, subtract the *x*-coordinates. (If the value is negative, find the absolute value.)

Geometry: 6.G.4

"Solve real-world and mathematical problems involving area, surface area, and volume."

4. "Represent three-dimensional figures using nets made up of rectangles and triangles, and use the nets to find the surface area of these figures. Apply these techniques in the context of solving real-world and mathematical problems."

BACKGROUND

A *net* is a two-dimensional drawing of connected polygons and/or circles in a plane that can be folded along adjacent sides to form a polyhedron.

A *polyhedron* is a three-dimensional shape with surfaces called faces that are two-dimensional. Each face consists of a polygon and the interior of the polygon.

A *prism* is a polyhedron with two parallel faces, called bases, that are the same size and shape. Prisms are classified according to the shape of their bases. If the bases are triangular, for example, the prism is called a triangular prism. If the bases are rectangular, the prism is called a rectangular prism.

A *pyramid* is a polyhedron with one face, called a base, that is a polygon. The other faces are triangles with a common vertex. A pyramid is classified according to the shape of its bases.

The *Platonic solids* are three-dimensional figures whose faces are regular polygons. A regular polygon is a polygon that has congruent sides and congruent angles. The five Platonic solids are listed below:

- *Tetrahedron:* 4 faces are equilateral triangles.
- *Cube:* 6 faces are squares.
- *Octahedron:* 8 faces are equilateral triangles.
- *Icosahedron:* 20 faces are equilateral triangles.
- *Dodecahedron:* 12 faces are regular pentagons.

To find the surface area of a three-dimensional figure, students should find the sum of the areas of the faces and bases.

 ACTIVITY: FINDING THE SURFACE AREA OF NETS

This is a two-day activity. Students may work individually or in small groups. On the first day, students will watch a simulation of folding a net into a three-dimensional figure. They will also select a polyhedron that they would like to make. On the second day they will construct their polyhedron and find its surface area.

Card stock paper; scissors; rulers; glue sticks; computers with Internet access; digital projector.

Print the nets that students will use to form their models.

1. Explain that nets are two-dimensional figures that will form a three-dimensional figure when they are folded.

2. Project the simulations of nets being used to create figures from the Web site http://www.mathsnet.net/geometry/solid/nets.html. Students will be able to view nets being folded into various three-dimensional figures.

3. Instruct your students to go to the Web site http://www.korthalsaltes.com to view models of three-dimensional figures. Clicking on a model will allow them to view the model and scroll down to view its net.

4. Instruct students to work either individually or in small groups at their computers. They are to select a model they would like to construct from a net.

5. Allow time for your students to select a figure they would like to make, limiting them to the Platonic solids, pyramids, and prisms. Ask them to write their selections on a sheet of paper with their name on it at the end of the class. If necessary, offer explanations about the types of figures listed in the Background for this activity.

6. After class, go to the Web site http://www.korthalsaltes.com and print your students' selections of nets on card stock paper.

7. On the second day, distribute materials—scissors, rulers, and glue sticks—and students' nets. Explain that the tabs on the nets are not a face of the figure, but are used to be glued and will hold the model together.

8. Explain that students are to find the area of each polygon that is a face or a base of the model. If necessary, review the formulas for finding the areas of squares ($A = s^2$); rectangles ($A = l \times w$); triangles $\left(A = \frac{1}{2}bh \right)$; and pentagons (divide the pentagon into five triangles, find the area of each, and find the sum of the areas). Emphasize that the sum of the areas of the faces and base(s) is equal to the surface area of the polyhedron.

9. Instruct your students to write their name on their model, find the area of each face and base, and record the areas on the model. Then they are to find the sum of the areas, and write the surface area on the paper on which they originally wrote their selection of the net.

Ask your students to brainstorm situations where finding the surface area is necessary in everyday life. (Answers might include the following: wrapping a package, painting the walls and ceiling of a room, or the amount of paint to use on products.)

Statistics and Probability: 6.SP.1

"Develop understanding of statistical variability."

1. "Recognize a statistical question as one that anticipates variability in the data related to the question and accounts for it in the answers."

BACKGROUND

Statistics is a branch of mathematics that focuses on the collection, organization, and analysis of data. In collecting statistical data, care must be taken to obtain thorough and accurate information. Recognizing the difference between a statistical question—a question to which the answers making up the data are expected to vary—and a nonstatistical question is fundamental to efficient collection of data.

 ACTIVITY: STATISTICAL QUESTIONS VERSUS NONSTATISTICAL QUESTIONS

Working in groups of four or five, students will generate examples of statistical and nonstatistical questions. They will present their questions to the class, providing explanations as to why their questions are statistical or not.

MATERIALS

One transparency per group; markers; erasers.

PROCEDURE

1. Explain to your students that a statistical question is a question designed to collect data that expects the data related to it to vary and takes account of the variability in its answer. This is unlike questions that have specific answers.

2. Offer these examples of statistical questions:

   * How tall are the members of the high school basketball team?

   * How old are the students in the school band?

   * What are the ethnic backgrounds of the students in my school?

   Note that in each case, the question anticipates various answers.

3. Offer these examples of nonstatistical questions:

   - Whose portrait is on a one-dollar bill?

   - Who is the president of the United States?

   - What is today's date?

   Note that in each case, a specific answer is anticipated.

4. Instruct your students to work in their groups to brainstorm and generate examples of statistical questions and nonstatistical questions. They are to choose one statistical question and write it neatly on their transparency. They should then choose one of the nonstatistical questions they generated and write it beneath their statistical question. They should be prepared to explain why these are, in fact, examples of a statistical and nonstatistical question.

5. Have each group present their questions to the class via the overhead projector. They should support their selection of these questions.

CLOSURE

Instruct your students to write a brief summary of the difference between a statistical question and a nonstatistical question.

Statistics and Probability: 6.SP.2

"Develop understanding of statistical variability."

2. "Understand that a set of data collected to answer a statistical question has a distribution which can be described by its center, spread, and overall shape."

BACKGROUND

A statistical question is one that is expected to have variability in the data related to it and accounts for this variability in its answers. For example, "How far do you live from your school?" is a statistical question because students will no doubt live various distances from their school. Another example of a statistical question is "What is your favorite ice cream?" Respondents to the question will likely have various types of favorite ice cream. The distribution of the answers to a statistical question can be described by its center (the mean, median, and mode), its spread (how far data is from the center), and overall shape (the shape of the graph of the data).

 ## ACTIVITY: AND THE ANSWER IS...

This activity requires two days. Working in pairs or groups of three, students will write a statistical question that they will pose to ten other students in the class. They will use their data to describe the center, spread, and overall shape of the distribution.

MATERIALS

Rulers.

PREPARATION

A day before you present this activity, ask each pair or group of students to write a statistical question that they wish to ask other students. Collect the questions to check that they are, in fact, statistical questions and that the questions are appropriate.

PROCEDURE

1. On the first day, instruct each pair or group of students to write a statistical question that they will later ask ten other students. If necessary, provide examples and counter-examples of statistical questions. Collect the questions and make sure that they are written correctly. (Any incorrect questions should be corrected before the next part of the activity.)

2. On the second day, hand back the statistical questions your students wrote. Instruct the pairs or groups to ask their question of ten other students in the class.

3. Instruct students to describe the data they collected in the following manner:

- *By its center.* Students may use the mean, median, or mode. Note that the mean is the average number of the data. The median is the middle value of the data, arranged in ascending or descending order. In this case, because there is an even number of data, the median is the average of the two middle values. The mode is the number that occurs most often in the data, and is particularly useful when students must identify favorites or the most popular.

- *By its spread.* Students may use the range, which is the highest value minus the lowest value. The spread may also be described as the values being close to the center or far apart.

- *By its shape.* Students may sketch a dot plot to describe the shape of the data.

4. If necessary, review the steps for sketching a dot plot. Note that students should do the following:

- Record each item of data on a number line.

- Place each dot above the number line to represent the frequency of the data points.

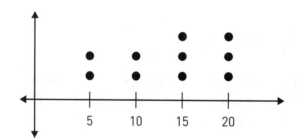

This dot plot represents 5, 5, 10, 10, 15, 15, 15, 20, 20, 20.

CLOSURE

Have your students share their results with the class. Ask and discuss the following question: Did any group pose a statistical question that could not be analyzed by its center, range, or shape? (The answer is no, because all data gathered from statistical questions can be analyzed in this manner.)

Statistics and Probability: 6.SP.3

"Develop understanding of statistical variability."

> 3. "Recognize that a measure of center for a numerical data set summarizes all of its values with a single number, while a measure of variation describes how its values vary with a single number."

BACKGROUND

A measure of center (mean, median, and mode) summarizes the values of a set of data with a single number. A measure of variation describes how spread out or scattered the set of data is. It includes the range and mean absolute deviation.

 ACTIVITY: MEASURE OF CENTER VERSUS MEASURE OF VARIATION

Working in pairs or groups of three, students will be given three sets of data. They will also be given cards identifying various measures of center and various measures of variation. They are to select which measures match their data.

MATERIALS

Scissors; glue sticks; reproducible, "Data Sets."

PROCEDURE

1. Distribute one copy of the reproducible to each pair or group of students. Explain that the reproducible contains three Data Sets. Eighteen cards are located at the bottom of the sheet, each representing either a measure of center or a measure of variation.

2. Explain that students are to find the mean, median, mode, range, and absolute mean deviation for each Data Set. If necessary, review the process for finding these measures, particularly how to find the mean absolute deviation, which is found by finding the mean and then finding the distance between each value and the mean. The sum of the distances divided by the number of data values is the mean absolute deviation. It is a measure of how far, on average, each value is from the mean.

3. Explain that students are to cut out the cards found at the bottom of the sheet, then match and glue the cards beneath the appropriate Data Sets, starting with measures of center and ending with measures of variation. Note that some cards will not be used.

Provide the answers to your students.

ANSWERS

The values are listed in this order: mean, median, mode, range, mean absolute deviation.
Data Set 1: 70, 67.5, 65, 40, 9; Data Set 2: 85, 90, 90, 35, 10; Data Set 3: 75, 70, 70, 60, $14\frac{2}{7}$

DATA SETS

Data Set 1 80, 70, 60, 50, 90, 70, 65, 65, 65, 85	
Measures of Center	Measures of Variation
(Place matching cards here)	
Data Set 2 80, 70, 95, 90, 90, 65, 90, 100	
Measures of Center	Measures of Variation
(Place matching cards here)	
Data Set 3 40, 90, 85, 100, 70, 70, 70	
Measures of Center	Measures of Variation
(Place matching cards here)	

The mean is 60	The median is 90	The median is 67.5	The range is 40	The mean is 70	The mean absolute deviation is 9
The median is 70	The mode is 65	The range is 60	The mean is 87	The mode is 70	The mean absolute deviation is $14\frac{2}{7}$
The mean is 85	The range is 35	The median is 76	The mode is 90	The mean is 75	The mean absolute deviation is 10

Statistics and Probability: 6.SP.4

"Summarize and describe distributions."

4. "Display numerical data in plots on a number line, including dot plots, histograms, and box plots."

BACKGROUND

Data can be organized and displayed in a variety of ways. A dot plot, histogram, and box plot are three common graphs used for summarizing and displaying data.

 ACTIVITY: CREATING DATA DISPLAYS

Working in groups of three or four, students will gather and display data in the form of a dot plot, histogram, and box plot.

MATERIALS

Rulers; unlined paper; reproducible, "Displaying Data."

PREPARATION

A few days before the activity, instruct your students to ask at least five adults (family members or friends) about how much time they need to travel to work each day. The answers students obtain will be the data they use for this activity.

PROCEDURE

1. Explain that each group of students is to combine the data they obtained and use the data to create a dot plot, histogram, and box plot. (For example, a group of four should have at least 20 items of data.)

2. Provide the following guidelines for organizing their data:

   • Express all times in minutes.

   • Round each time to the nearest multiple of 5.

   • Arrange data from the least to the greatest.

3. Distribute and review the information on the reproducible with your students. Note that the reproducible contains a description of each graph and instructions for constructing the graphs.

4. Remind your students that their graphs should be neat, accurate, and labeled correctly. ·

CLOSURE

Ask your students the following questions: Which graph did they find easiest to construct? Why? Which was hardest? Why?

(Note: Instruct your students to retain their data and graphs, which will be used in the next activity.)

DISPLAYING DATA

Dot plots, histograms, and box plots are three ways data can be displayed. Use the data the members of your group obtained to create these graphs.

A DOT PLOT

Also called a line plot, a dot plot represents a set of data by using dots placed over a number line. To make a dot plot, do the following:

1. Draw a number line, labeling the numbers. (For your dot plot in this activity, label the numbers in multiples of 5.)

2. Place a dot above the number line each time a number representing data occurs.

A HISTOGRAM

A special type of bar graph, a histogram displays the frequency of data that has been organized in equal distributions. To make a histogram, do the following:

1. Determine the number of bars you will include on your graph.

2. Subtract the smallest value of the data from the largest value of the data, and divide that answer by the number of bars to obtain the bar width.

3. Draw horizontal and vertical axes. Label the horizontal axis with the data and the vertical axis with the frequency.

4. Draw each bar (with no spaces between them), using the frequency as the height.

A BOX PLOT

Also called a box-and-whisker plot, a box plot displays the median of a set of data, a median of each half of the data, and the least and greatest values of the data. To make a box plot, to the following:

1. Find the median of the data. (This divides the data into two parts.)

2. Find the median of the first half of the data.

3. Find the median of the second half of the data.

4. Draw a box using the median of the first half of the data as the left end, and the median of the second half of the data as the right end. Mark the median of the data with a verical line.

5. Draw a line from the box to the smallest number.

6. Draw a line from the box to the largest number.

Statistics and Probability: 6.SP.5

"Summarize and describe distributions."

5. "Summarize numerical data sets in relation to their context, such as by:

   a. "Reporting the number of observations.

   b. "Describing the nature of the attribute under investigation, including how it was measured and its unit of measurement.

   c. "Giving quantitative measures of center (median and/or mean) and variability (interquartile range and/or mean absolute deviation), as well as describing any overall pattern and any striking deviations from the overall pattern with reference to the context in which the data were gathered.

   d. "Relating the choice of measures of center and variability to the shape of the data distribution and the context in which the data were gathered."

BACKGROUND

Analyzing data is essential for data display. An explanation of how the data was collected, the number of data, how accurate the data is, how the measures of central tendencies and/or variability are used, the identification of outliers, and the identification of patterns can enhance any numerical display. (Note that for this activity, students will work with the same data they obtained for the previous activity, "Creating Data Displays," for 6.SP.4.)

ACTIVITY: SUMMARIZING DATA

Working in groups of three or four (preferably the same groups students worked in for the prior activity), students will write a description of the data they previously obtained and displayed.

MATERIALS

Data regarding travel time to work; reproducible, "Guidelines for Summarizing Data."

PROCEDURE

1. Instruct your students that they are to use the data they obtained previously. They may also use their data displays (from the previous activity) for reference.

2. Explain that each group is to write a summary of their data.

3. Distribute copies of the reproducible. Explain that it is a reference sheet that includes the topics students should include in their summaries. It also includes definitions and how to determine the values they should include in their summaries. Discuss the definitions and processes, offering any necessary examples.

CLOSURE

Exhibit students' displays and discuss their summaries.

GUIDELINES FOR SUMMARIZING DATA

Your group is to write a summary of the data you collected about the traveling time of adults going to work. You may use any dot plot, histogram, and box plot you created previously to support your summary.

Your summary should include answers to the following questions:

1. How much data did your group collect?

2. How did you collect your data?

3. Was the data exact?

4. Are there any values that are unusually large or small compared with the rest of the values? (These are called outliers.)

5. How close is your data to its center? This can be found by using measures of central tendencies: the mean, median, and mode.

   - The mean is an average and can be found by adding all of the data and dividing the sum by the number of data. It is not a good measure to use if there are outliers, because outliers can have a major effect on the mean.

   - The median is the middle number of the data when the numbers are arranged in order from the least to the greatest (or the greatest to the least). The median is less affected by outliers than the mean and mode.

   - The mode is the number of the data that occurs most often.

6. How spread out is your data? This can be found by using measures of variability, the range, the interquartile range, and/or mean absolute deviation.

   - The range of the data is the highest value minus the lowest value.

   - The interquartile range is the middle half of the data and tells how spread out the middle values are.

   - The mean absolute deviation (also known as the mean deviation) uses each piece of data in its computation to find out how far on average the data deviates from the mean. To calculate the mean absolute deviation, first find the mean. Then subtract the mean from each piece of data. Find the absolute value of each difference. Add the total of the absolute values and divide the sum by the number of data.

7. Do you notice any patterns? If yes, what are they? What might they be a result of?

8. How do the measures of central tendencies and variation relate to the shape of your displays?

Standards and Activities for Grade 7

Ratios and Proportional Relationships: 7.RP.1

"Analyze proportional relationships and use them to solve real-world and mathematical problems."

> 1. "Compute unit rates associated with ratios of fractions, including ratios of lengths, areas and other quantities measured in like or different units."

BACKGROUND

A rate is a ratio that is used to compare different types of quantities. A unit rate describes how many of the first quantity corresponds to one unit of the second quantity. Some common unit rates include miles per hour, miles per gallon, dollars per hour, and meters per second. In each example the second unit is 1—one hour, one gallon, one hour, and one second.

To find a unit rate, scale the denominator of the original ratio so that 1 is the denominator. For example, if a person can walk 15 meters in 20 seconds, the ratio of meters to seconds is $\frac{15}{20}$. To find the unit rate, divide the numerator and denominator by 20 so that the denominator is 1. The unit rate for this ratio is 0.75 meters per second.

The process for finding unit rates associated with fractions is the same. For example, if a person can walk $\frac{4}{5}$ meter in $\frac{1}{2}$ second, the ratio of meters to seconds can be expressed as a complex fraction, $\frac{\frac{4}{5}}{\frac{1}{2}}$. A complex fraction is a fraction in which the numerator, denominator, or numerator and denominator are fractions. To convert the ratio based on a complex fraction to a unit rate, multiply the numerator and denominator by the reciprocal of the denominator, so that the denominator is 1. The unit rate for the example above is $1\frac{3}{5}$ meters per second.

$$\frac{\frac{4}{5} \times \frac{2}{1} = \frac{8}{5} = \frac{8}{5}}{\frac{1}{2} \times \frac{2}{1} = \frac{2}{2} = 1} = \frac{8}{5} = 1\frac{3}{5}$$

ACTIVITY: WHAT IS THE UNIT RATE?

For this activity, you will give half of your students a card that contains a ratio of fractions that they will convert to a unit rate. You will give the other half of the class a card that contains the unit rate. Students must match ratios and unit rates.

MATERIALS

2-page reproducible, "Equivalent Rates."

PREPARATION

Make a copy of the reproducible and cut out each box. (The original will serve as your key.) You should have a total of 30 cards—15 of which are ratios and 15 that are unit rates.

PROCEDURE

1. Mix up the cards and have your students select a card from a hat or box.

2. Explain that some students will have a card with a ratio that must be converted to a unit rate. Other students will have the corresponding unit rate. Their goal is to find the ratio and unit rate that are equivalent.

3. Give students a few minutes to think about their cards. Those who have a card with a ratio should try to determine the unit rate. Those who have the unit rate should think about ratios that could be equivalent to their unit rate.

4. After a few minutes, instruct the students who have cards with the ratios to walk around the room and check with students at their seats who have cards with unit rates. They are to match ratios with equivalent unit rates. Students who have a matching unit rate and ratio should sit together.

5. Once students have matched ratios and unit rates, discuss the process with the class, focusing on how students determined the correct matches.

6. You may repeat this activity as many times as you feel necessary.

CLOSURE

Write this question on the board: "What is the process for finding a unit rate when you have a ratio with fractions in the numerator and denominator?" Allow your students about five minutes to reflect upon and write about this process. Select a few students to read their answers to the class.

EQUIVALENT RATES

Ratios	Unit Rates
$\dfrac{\frac{3}{4} \text{ cup of flour}}{\frac{1}{2} \text{ teaspoon of oil}}$	$1\frac{1}{2}$ cups of flour per 1 teaspoon of oil
$\dfrac{\frac{1}{2} \text{ cup of flour}}{\frac{3}{4} \text{ teaspoon of oil}}$	$\frac{2}{3}$ cup of flour per 1 teaspoon of oil
$\dfrac{1\frac{1}{2} \text{ cups of flour}}{2 \text{ teaspoons of oil}}$	$\frac{3}{4}$ cup of flour per 1 teaspoon of oil
$\dfrac{\frac{4}{5} \text{ cup of flour}}{\frac{1}{4} \text{ teaspoon of oil}}$	$3\frac{1}{5}$ cups of flour per 1 teaspoon of oil
$\dfrac{2\frac{1}{3} \text{ cups of flour}}{1\frac{1}{4} \text{ teaspoons of oil}}$	$1\frac{13}{15}$ cups of flour per 1 teaspoon of oil
$\dfrac{\frac{3}{4} \text{ mile}}{\frac{1}{4} \text{ hour}}$	3 miles per hour
$\dfrac{\frac{8}{9} \text{ mile}}{\frac{9}{10} \text{ hour}}$	$\frac{80}{81}$ mile per hour
$\dfrac{\frac{1}{3} \text{ mile}}{\frac{4}{5} \text{ hour}}$	$\frac{5}{12}$ mile per hour

(continued)

Ratios	Unit Rates
$\dfrac{1\frac{1}{2} \text{ miles}}{3\frac{1}{4} \text{ hours}}$	$\dfrac{6}{13}$ mile per hour
$\dfrac{\frac{5}{6} \text{ mile}}{1\frac{1}{4} \text{ hours}}$	$\dfrac{2}{3}$ mile per hour
$\dfrac{1\frac{1}{10} \text{ dollars}}{2 \text{ items}}$	$\dfrac{11}{20}$ dollar per item
$\dfrac{2\frac{2}{5} \text{ dollars}}{2 \text{ items}}$	$1\frac{1}{5}$ dollars per item
$\dfrac{3\frac{1}{2} \text{ dollars}}{2 \text{ items}}$	$1\frac{3}{4}$ dollars per item
$\dfrac{3 \text{ dollars}}{2 \text{ items}}$	$1\frac{1}{2}$ dollars per item
$\dfrac{2\frac{1}{2} \text{ dollars}}{5 \text{ items}}$	$\dfrac{1}{2}$ dollar per item

Ratios and Proportional Relationships: 7.RP.2

"Analyze proportional relationships and use them to solve real-world and mathematical problems."

2. "Recognize and represent proportional relationships between quantities.

   a. "Decide whether two quantities are in a proportional relationship, e .g., by testing for equivalent ratios in a table or graphing on a coordinate plane and observing whether the graph is a straight line through the origin.

   b. "Identify the constant of proportionality (unit rate) in tables, graphs, equations, diagrams, and verbal descriptions of proportional relationships.

   c. "Represent proportional relationships by equations.

   d. "Explain what a point (x, y) on the graph of a proportional relationship means in terms of the situation, with special attention to the points $(0, 0)$ and $(1, r)$ where r is the unit rate."

BACKGROUND

Two quantities that are in a proportional relationship represent a linear function and can be written as $y = mx$, because there is a constant rate of change in the x-variable that corresponds to a constant rate of change in the y-variable. m is the constant of proportionality. When this relationship is graphed, the ordered pair $(1, r)$ can be converted to the unit rate $\frac{r}{1}$ because the value of x is 1.

ACTIVITY: PROPORTIONS SCAVENGER HUNT

Students will look through various sources to find examples of ratios. They will convert each ratio into its unit rate. They will then use the unit rate to create tables and graphs, and then write an equation to show the relationship between the quantities in the ratio.

MATERIALS

Math texts; reference books; newspapers; magazines; rulers; graph paper. Optional: computers with Internet access.

PROCEDURE

1. For a homework assignment the day before you present this activity in class, instruct your students to find three ratios. They may look in newspapers, books, or magazines, or online.

2. The next day, explain that students will be using these ratios to investigate proportions. Instruct them to create a table. Depending on the background knowledge of your students, you may have to provide an example. The table should have the labels of each ratio on top (for example, miles and hours), followed by the ratio the students have found. Instruct your students to convert the ratios they found into the unit rates. Students can multiply or divide the unit rate by 1 which can be expressed as $\frac{2}{2}$, $\frac{3}{3}$, and so on to find equivalent ratios, or proportions. Emphasize to your students that a proportion states that two ratios are equal. Using the example of miles per hour, show them how to set up and complete a table. You may want them to complete this table for ten values.

3. Once students have finished their table, based on their ratios, instruct them to graph their data in a coordinate plane.

4. Instruct them to write a summary of what they noticed from the ratio, the unit rate, table, and graph. For example, they should consider the shape of the graph. (It should be a line through the origin.) They should examine the unit rate of the ratio as well as how the numbers in the table and graph are increasing or decreasing. When your students discover the unit rate, they should label it in the table and the graph.

5. After your students have examined the table and graph, challenge them to find an equation for each ratio. Remind them that the y-value is the output of the equation. This equation must be true for every value in their table and graph.

6. After students are finished with this activity, instruct them to select one of their ratios to present to the class. As they present their ratios, they should identify the original ratio, unit rate, table, graph, and equation.

CLOSURE

Instruct your students to discuss with a partner how ratios, proportions, tables, graphs, equations, and unit rates are related. Then discuss the answers as a class. Students should realize that a ratio may be a proportion. Emphasize that if it is a proportion, it represents a linear function whose graph forms a line that intersects the axes at the origin.

Ratios and Proportional Relationships: 7.RP.3

"Analyze proportional relationships and use them to solve real-world and mathematical problems."

> 3. "Use proportional relationships to solve multistep ratio and percent problems."

BACKGROUND

Percent is a ratio that represents a part of 100. A proportion is a statement that two ratios are equal.

To find a percent, first write the quantities as a ratio. For example, a student correctly answered 20 out of 25 questions on a test. To find the percent of correct answers, first write the ratio of the number of correct answers to the total number of questions, $\frac{20}{25}$. Then write a proportion in which the denominator of the second fraction is 100, $\frac{20}{25} = \frac{n}{100}$. Solve for n. $n = 80$. Because $\frac{80}{100} = 80\%$, the student answered 80% of the questions correctly.

 ## ACTIVITY 1: GIFTS FOR THE HOLIDAYS

Working in pairs or groups of three, students will use percents and discounts to purchase $1,000 worth of gifts for the holidays. They will search through catalogs and sales circulars to find items to purchase.

MATERIALS

Sales circulars; catalogs; reproducible, "Gift Ordering Form." Optional: computers with Internet access; calculators.

PREPARATION

Collect sales circulars and catalogs a few days before you begin this activity. You might ask students to bring some from home to class.

PROCEDURE

1. Explain to your students that they have $1,000 to spend on gifts for the holidays. Students should look through sales circulars and catalogs to determine which items they would like to purchase. You can expand their sources by encouraging them to

check sales material online at store Web sites. They must try to spend all of their money without going over budget.

2. Review the reproducible, "Gift Ordering Form," with your students. Explain what each section means.

- Store — the sales circular, catalog, or online Web site from which they are purchasing the item.

- Item — the product they are purchasing.

- Quantity — the number of specific items they are purchasing.

- Original price — the list price.

- Discount percent — the percent off the original price the store is offering. (Some stores may provide this information, whereas other stores may simply have the comparison price. If the discount percent is not provided, students must subtract the sales price from the original price to find the discount. They should write and solve a proportion: $\dfrac{discount}{original\ price} = \dfrac{n}{100}$.)

- Sale price — price of the item on sale.

- Total — total cost for that item (multiply the quantity by the sale price).

- Cumulative total — the total cost for all the items purchased up to that point.

- Subtotal — the total cost of all the items purchased before tax or shipping.

- Tax — the amount of tax, 7%, that will be added to the subtotal.

- Shipping — this fee is 3% of the total after tax.

- Total — the total cost, including tax and shipping. (If necessary, explain how students should compute tax and shipping costs.)

3. Encourage your students to be savvy shoppers by selecting the best gifts at the best prices.

CLOSURE

Instruct your students to exchange ordering forms with another pair or group of students. Each should check the ordering form that they received and highlight any answers they feel are incorrect. Students should then conference with each other, discuss any incorrect answers, and help each other correct the work.

Name _____ Date _____

GIFT ORDERING FORM

Store	Item	Quantity	Original Price	Discount Percent	Sale Price	Total	Cumulative Total

Subtotal _____

Tax (7%) _____

Shipping _____

(3% of total after tax)

Total _____

(including tax and shipping)

ACTIVITY 2: VERY INTERESTING

Working in pairs or groups of three, students will use the simple interest formula to compute interest. They will look for patterns and use proportional reasoning to select the appropriate numbers to earn a specified amount of interest.

OPTIONAL MATERIALS

Calculators.

PROCEDURE

1. Introduce the simple interest formula to your students. $I = Prt$. Explain that simple interest is used to determine the interest paid on money that is borrowed or interest earned on money that is invested.

2. Explain what each variable represents.

   - I is the interest paid or earned.

   - P is the principal amount borrowed or invested.

   - r is the rate of interest.

   - t is the amount of time (in years) the interest is calculated for.

3. Explain that students will be given five principals, five rates of interest, five time periods, and five amounts of interest. They are to select the appropriate principal, rate, and time period that will pay a specific amount of interest.

4. Present these values to your students:

Principal	Rate of Interest	Time
$1,000	3%	1 year
$500	1%	6 months
$750	2%	2 years
$800	2.5%	3 years
$150	1.5%	1.5 years

5. Explain that students should substitute the above values in the interest formula to match the following interest amounts: $40, $7.50, $22.50, $6.75, and $10.

CLOSURE

Summarize the activity and discuss any problems that were particularly difficult.

Answer Key: $1,000 \times 0.01 \times 1 = \$10$; $500 \times 0.03 \times 1.5 = \$22.50$; $750 \times 0.02 \times 0.5 = \$7.50$; $800 \times 0.025 \times 2 = \$40$; $150 \times 0.015 \times 3 = \$6.75$

The Number System: 7.NS.1

"Apply and extend previous understandings of operations with fractions to add, subtract, multiply, and divide rational numbers."

1. "Apply and extend previous understandings of addition and subtraction to add and subtract rational numbers; represent addition and subtraction on a horizontal or vertical number line diagram.

   a. "Describe situations in which opposite quantities combine to make 0.

   b. "Understand $p + q$ as the number located a distance $|q|$ from p, in the positive or negative direction depending on whether q is positive or negative. Show that a number and its opposite have a sum of 0 (are additive inverses). Interpret sums of rational numbers by describing real-world contexts.

   c. "Understand subtraction of rational numbers as adding the additive inverse, $p - q = p + (-q)$. Show that the distance between two rational numbers on the number line is the absolute value of their difference, and apply this principle in real-world contexts.

   d. "Apply properties of operations as strategies to add and subtract rational numbers."

BACKGROUND

Every number can be graphed on a number line. When adding rational numbers, students may use the number line to move in the positive direction (right) or the negative direction (left) depending upon the numbers they are adding. For example, to find the sum of $5 + (-3)$, students must start at 5 on the number line. Because -3 is a negative number, students will move three spaces left on the number line to arrive at 2.

When subtracting rational numbers, students should use the additive inverse. The additive inverse is defined as the opposite of a number. For example, the additive inverse of 7 is -7. The additive inverse of -2 is 2. To subtract two numbers, write the problem by adding the opposite (or additive inverse) of the second number. For example, $3 - (-7) = 3 + 7$ or 10, and $3 - 9 = 3 + (-9)$ or -6. The absolute value, shown by the symbol $|x|$ of a number, is the number's distance from 0 on the number line. The absolute value of a number will always be positive, except the absolute value of 0 which is 0. The distance between two points on a number line is the absolute value of their difference. For example, the difference between 3 and -9 is $|3 - (-9)|$ or 12.

 **ACTIVITY: THE VIRTUAL CLASSROOM**

Working individually or with a partner, students will learn concepts of absolute value and additive inverse from online interactive Web sites. They will then use these concepts to practice addition and subtraction of rational numbers.

Computers with Internet access.

1. Provide your students with the following Web sites:

   • *Absolute Value:* http://www.kidsknowit.com/interactive-educational-movies/free-online-movies.php?movie = Absolute%20Values. Students should click on "Play" to watch the interactive movie about absolute values.

   • *Additive Inverse:* http://www.curriki.org/xwiki/bin/view/Coll_wincurriki/Inverse OperationandAbsolutevalueproperties?bc=;Coll_wincurriki.Winpossible. Students should scroll down to the mini-lesson on additive inverse. You may encourage your students to listen to other mini-lessons at this site as well.

   • *Practice Problems:* http://www.ixl.com/math/grade-7/add-and-subtract-rational-numbers. Note that for fractions students will have to submit their answers with a diagonal fraction bar. For example, a simple fraction would be submitted as 2/5, and a mixed number would be submitted as 4 1/3. Remind your students to include negative signs as necessary. Encourage them to listen to the explanation for problems they have trouble solving.

2. Students may work with a partner when visiting the Web sites. Encourage them to take notes.

3. When working on the practice problems, students should rewrite each problem and show all their work. This will serve as a reference should they have questions regarding any of the problems.

4. After students finish their work at all three Web sites, select a few students to choose a problem and write it on the board, along with their work. Discuss the procedures each student went through to answer the problems.

CLOSURE

Instruct your students to write a paragraph explaining the procedures for adding and subtracting rational numbers. Ask them to include examples when they would use these skills in the real world, or what real-world situations reflect these skills.

The Number System: 7.NS.2

"Apply and extend previous understandings of operations with fractions to add, subtract, multiply, and divide rational numbers."

2. "Apply and extend previous understandings of multiplication and division and of fractions to multiply and divide rational numbers.

   a. "Understand that multiplication is extended from fractions to rational numbers by requiring that operations continue to satisfy the proprieties of operations, particularly the distributive property, leading to products such as $(-1)(-1) = 1$ and the rules for multiplying signed numbers. Interpret products of rational numbers by describing real-world contexts.

   b. "Understand that integers can be divided, provided that the divisor is not zero, and every quotient of integers (with non-zero divisor) is a rational number. If p and q are integers, then $-(p/q) = (-p)/q = p/(-q)$. Interpret quotients of rational numbers by describing real-world contexts.

   c. "Apply properties of operations as strategies to multiply and divide rational numbers.

   d. "Convert a rational number to a decimal using long division; know that the decimal form of a rational number terminates in 0s or eventually repeats."

BACKGROUND

The rules for multiplying or dividing two rational numbers are listed below:

- If the signs are the same, the answer is always positive.
- If the signs are different, the answer is always negative.
- If a number is multiplied by zero, the product is zero.
- Division by zero is undefined.
- If zero is divided by any rational number, the quotient is zero.

When multiplying or dividing more than two numbers, follow these rules:

- When multiplying or dividing any number of positive rational numbers, the answer is positive.
- When multiplying or dividing an even number of negative rational numbers, the answer is positive.
- When multiplying or dividing an odd number of negative rational numbers, the answer is negative.

Understanding these rules will help students multiply and divide rational numbers proficiently. Students should also know and be able to apply the distributive property, which states that $a(b + c) = ab + ac$ and $(b + c)a = ba + ca$. The distributive property facilitates calculations because it allows students to multiply a sum by multiplying the addends separately and then adding or subtracting the products as stated in the problem. For example, $5 \times 3\frac{1}{5}$ can be rewritten as $5\left(3 + \frac{1}{5}\right)$. $5 \times 3 = 15$ and $5 \times \frac{1}{5} = 1$. The sum is 16.

 ## ACTIVITY 1: THE DISTRIBUTIVE PROPERTY WAR

Pairs of students will play a card game similar to the game "War." Each student will receive ten cards, each of which contains a problem they must solve by applying the distributive property. The student who obtains the most cards at the end of the hand wins the game.

MATERIALS

One pair of scissors for each pair of students; reproducible, "The Distributive Property War Cards." Optional: standard index cards.

PREPARATION

Make enough copies of the reproducible so that each pair of students has one copy. (If you have an odd number of students in class, you may make another set of cards and allow three to play against each other. Of course, depending on your students, you may prefer to make sets of cards with different problems.)

PROCEDURE

1. Explain the distributive property to your students and provide several examples using rational numbers. Emphasize that the distributive property allows the distribution of numbers to other numbers by using multiplication. It allows sums to be separated and then multiplied.

2. Students will work in pairs to play the card game. Explain that the game is similar to the card game "War." Each student plays with ten cards. They will each flip one card over, then compute the answer to the problem shown on their card. The student whose answer has the higher value will take both cards. They will repeat this process until they have completed the problems on all of the cards. The player who has more cards at the end of the hand is the winner. In the case of a draw, you should provide a problem that both students must solve. The student who solves the problem first is the winner.

3. Pass out copies of the reproducible, one to each pair of students. Instruct your students to cut out the cards. Each student takes ten cards, randomly mixes them up, and places them face down. Students are to flip a card over, one at a time, and solve the problem. Instruct them to write out each problem on a separate sheet of paper so that they can check each other's work. They must verify that the answers to the problems are correct. The student whose problem has the higher value takes both cards. This process is followed for the next nine cards. At the end of the hand, the student who has more cards is the winner.

4. When your students have finished the game, read the answers to the problems so that students can check that they, in fact, found the correct answers for all the cards.

5. You can extend this activity by having each student make a set of ten cards (writing one problem per card) on index cards. You can then collect the cards, shuffle them, and redistribute them to your students so they may play again with different problems.

CLOSURE

Instruct your students to write a definition of the distributive property in their own words. Ask them to answer the following question: Where and why would you use the distributive property? After students write their responses, encourage them to share their answers with a partner. Then select a few students to read their answers aloud.

ANSWERS TO THE PROBLEMS ON THE CARDS

Column 1: 235, −192, 48, −104, 152, 154, 8, 14.5, 13, −36

Column 2: 237, 4, −190, −588, 171, −42, 5, 9, $4\frac{1}{2}$, −12

$5(40 + 7)$	$3(70 + 9)$
$-4(40 + 8)$	$2(10 - 8)$
$2(30 - 6)$	$-5(30 + 8)$
$-8(10 + 3)$	$-7(80 + 4)$
$2(80 - 4)$	$9(10 + 9)$
$7(2 + 20)$	$6(3 - 10)$
$\frac{1}{2}(10 + 6)$	$\frac{1}{5}(20 + 5)$
$0.25(50 + 8)$	$0.5(8 + 10)$
$\frac{1}{3}(9 + 30)$	$\frac{1}{4}(16 + 2)$
$-6(-4 + 10)$	$-4(-5 + 8)$

ACTIVITY 2: WHICH ONE SHOULD BE EXCLUDED?

For this activity, students may work individually or with a partner. They are to determine which expression in a set of three is not equivalent to the other two.

MATERIALS

Reproducible, "This One Does Not Belong."

PROCEDURE

1. Review the directions on the reproducible with your students. For each row, students must circle the expression that is not equivalent to the others and explain why.

2. Remind them to formulate their explanations clearly.

CLOSURE

Discuss the answers to the problems with your students, focusing on their explanations. For more difficult problems, select a few students to show their work on the board.

ANSWERS

1) The second expression equals $\frac{1}{3}$; the others equal $-\frac{1}{3}$.

2) The second expression equals $-0.5\overline{3}$; the other expressions equal $0.5\overline{3}$.

3) The second expression equals -7; the others equal 7.

4) The third expression equals -0.2; the others equal -0.3.

5) The second expression equals $\frac{3}{14}$; the others equal $3\frac{3}{7}$.

6) The third expression equals $0.\overline{72}$; the others equal $-0.\overline{72}$.

7) The third expression equals 4.25; the others equal -4.25.

8) The third expression equals $\frac{-10}{13}$; the others equal -1.3.

9) The second expression equals 30; the others equal -30.

10) The second expression equals 3.6; the others equal -3.6.

Copyright © 2012 by Judith A. Muschla, Gary Robert Muschla, and Erin Muschla.

Name _____ Date _____

THIS ONE DOES NOT BELONG

Circle the expression that is not equivalent to the other expressions in each row. Show your work on another sheet of paper and explain why it is not equivalent to the other expressions.

1. $-\dfrac{1}{3}$ $0.\overline{3}$ $\dfrac{1}{-3}$

2. $\dfrac{2}{3} \times \dfrac{4}{5}$ $\dfrac{2}{3} \times \left(-\dfrac{4}{5}\right)$ $0.5\overline{3}$

3. $-42 \div (-6)$ $-42 \div 6$ $\dfrac{-42}{-6}$

4. $\dfrac{1}{4} \times \left(-1\dfrac{1}{5}\right)$ $0.25 \, (-1 - 0.2)$ $0.25 \, (-1 + 0.2)$

5. $-\dfrac{6}{7} \div \left(-\dfrac{1}{4}\right)$ $-\dfrac{6}{7} \div \left(-\dfrac{4}{1}\right)$ $-\dfrac{6}{7} \times \left(-\dfrac{4}{1}\right)$

6. $\dfrac{-8}{11}$ $-0.\overline{72}$ $-8 \div (-11)$

7. $(2.5)\,(-1.7)$ $\dfrac{5}{2} \times \left(-1\dfrac{7}{10}\right)$ $2\dfrac{1}{2} \times 1\dfrac{7}{10}$

8. $3\overline{)-3.9}$ $-1.3\,(1)$ $\dfrac{-3}{3.9}$

9. $2 \times (-3) \times 5$ $-10 \times (-3)$ $-100 \div 3\dfrac{1}{3}$

10. $\dfrac{4}{5} \times \left(-4\dfrac{1}{2}\right)$ $0.8 \times (4.5)$ $5 \div \left(-1\dfrac{7}{18}\right)$

The Number System: 7.NS.3

"Apply and extend previous understandings of operations with fractions to add, subtract, multiply, and divide rational numbers."

3. "Solve real-world and mathematical problems involving the four operations with rational numbers."

BACKGROUND

To solve real-world problems, students should be fluent in computing rational numbers. One real-world problem is selecting stocks that are likely to provide a positive return on invested money.

 ACTIVITY: PLAYING THE STOCK MARKET

Working first individually, then in groups of three or four, students will pretend to purchase stocks and calculate their gains or losses. They will analyze and summarize their investments.

MATERIALS

Newspapers and/or computers with Internet access; reproducible, "Stock Market Recording Sheet—Example;" reproducible, "Stock Market Recording Sheet;" reproducible, "Stock Market Worksheet."

PREPARATION

If you choose to use newspapers to track the daily performance of stocks, arrange to have enough copies for your students to use during the activity. You might ask colleagues to bring in copies of daily newspapers that contain stock quotes and financial information. Some students may also have access to newspapers at home. Make enough copies of the reproducibles so that each student has a copy of each one.

PROCEDURE

1. Explain the basic principles of the stock market to your students. A thorough summary of the stock market, stocks, and history can be found at the Web site http://library.thinkquest.org/3088/stockmarket/introduction.html. You may wish to summarize this for your students or have them visit this site.

2. Explain that for the first part of this activity, students will work individually. Ask them to imagine that you are "giving" them $1,000 to invest in the stock market. They can purchase stock in one company without exceeding $1,000. They should spend as much of their money as possible on the purchase of stock. Any remaining money will be placed in a bank account; however, doing so will result in no gain because of the short period of time (one week). Students will track their stock for seven business days in order to calculate their gains or losses.

3. Provide newspaper stock listings to your students or direct them to the Web site http://finance.yahoo.com/ where they can research companies, find stock prices, examine trends of prices, and track stocks.

4. After students select a company in which to invest, distribute copies of the reproducible, "Stock Market Recording Sheet—Example." The example contains information about the fictional Peach Computer Company. Review the information on the reproducible with your students and explain the terms and meaning of each column. Note that the first table provides information about the purchase of the stock. The second table includes:

   • The opening price is the price of one share of the stock at the beginning of the day.

   • The closing price is the price of one share of the stock when the stock market closes.

   • Change is the amount of difference for one share of stock in the opening and closing price. This change will be positive if the stock increased in value and negative if the stock decreased in value.

   • Money made or lost represents the amount of money gained or lost for the total number of stocks purchased. For example, ten shares of stock of the company were purchased. To calculate the money made or lost, students must multiply 10 times the change.

   • The net worth represents the total amount of the investment. It is found by multiplying the opening price by the number of shares purchased and adding the money made or lost.

5. Distribute copies of the reproducibles, "Stock Market Recording Sheet" and "Stock Market Worksheet." Instruct your students to complete the recording sheet and track their stocks for seven business days, filling out a row on the recording sheet each day. The worksheet provides students with guidance to calculate their net worth. This can be completed during school or at home.

6. After seven days, divide your students into small groups and instruct them to share their findings. They should discuss the questions, such as the following:

- Describe the performance of the stocks of your companies. Did your stocks go up or down? By how much? How much money did you gain or lose during the seven days? What was the percent change for your companies' stocks? Why is percent change helpful in making comparisons between companies' stocks?

- Explain how you calculated the amount of money you made or lost during the week. Explain how you calculated the percent change.

- Ask your students to make predictions by posing these questions: Would you choose these same stocks again? Why or why not? How do you think your company will perform during the next year? Will the stock go up or down? Why do you think this?

- Who gained the most money in your group? What might be the reason for this?

CLOSURE

Have each group report on one of the questions they answered. Also, discuss as a class things that students found interesting or surprising as they completed this activity.

STOCK MARKET RECORDING SHEET — EXAMPLE

Stock	Stock Ticker Symbol	Purchase Price	Number of Shares Purchased	Total Amount Spent	Balance
Peach Computer Company	PCC	$84.17	10	$841.70	$158.30

The amount of money placed in a bank account: $158.30

Two-Day Chart (for the Stock of the Peach Computer Company)

Day	Opening Price	Closing Price	Change (Positive or Negative)	Money Made or Lost	Net Worth
1	$84.17	$85.36	+1.19	$11.90	$853.60
2	$85.36	$83.80	−1.56	−$15.60	$838.00

Equations for determining net worth:

Day 1: ($84.17 × 10) + $11.90 = $853.60

Day 2: $853.60 + (−$15.60) = $838.00

Name _____   Period _____

STOCK MARKET RECORDING SHEET

Company Name: _____

Stock	Stock Ticker Symbol	Purchase Price	Number of Shares Purchased	Total Amount Spent	Balance

The amount of money placed in a bank account: _____

Day	Opening Price	Closing Price	Change (Positive or Negative)	Money Made or Lost	Net Worth
1					
2					
3					
4					
5					
6					
7					

Name _____ Date _____

STOCK MARKET WORKSHEET

Equations for Determining Net Worth:

Day 1: _____

Day 2: _____

Day 3: _____

Day 4: _____

Day 5: _____

Day 6: _____

Day 7: _____

How much money did you deposit in the bank? _____

Original net worth as of Day 1: _____

Final net worth as of Day 7: _____

Total change:

(Final net worth on Day 7) — (Original net worth on Day 1) = _____

Percent Change: _____

(Remember that to calculate "percent change" you must analyze the ratio of "Total Change" to "Original Net Worth on Day 1" and convert this ratio to a percent.)

Total net worth (including money in the bank): _____

Expressions and Equations: 7.EE.1

"Use properties of operations to generate equivalent expressions."

1. "Apply properties of operations as strategies to add, subtract, factor, and expand linear expressions with rational coefficients."

BACKGROUND

The basis of simplifying algebraic expressions by addition, subtraction, factoring, or expanding lies in combining like terms. Terms are parts of an algebraic expression that are separated by an addition or subtraction sign. The terms in the expression $2x + 6$ are $2x$ and 6. Sometimes expressions have terms that are alike, or similar. When two or more terms have the same variable raised to the same power, they are called *like terms*. Like terms differ only in their numerical coefficient. Combining like terms can be used to simplify expressions and make computation easier. For example, $3x + 5x + 4 = 8x + 4$ and $5y - 4y + 6 = y + 6$.

The distributive property, $a(b + c) = ab + ac$ and $(b + c)a = ba + ca$, can be used to factor and expand linear expressions. For example, to expand the expression $3(x + 5)$, you must first multiply $3 \cdot x$ and then add it to $3 \cdot 5$. The expanded expression is $3x + 15$. Conversely, each expanded expression can be factored by removing a common factor from both terms. For example, the terms of $5x - 20$ are $5x$ and 20. Both terms have a common factor of 5. 5 can be factored from $5x - 20$. $5x - 20 = 5(x - 4)$.

 ## ACTIVITY: ALGEBRAIC EXPRESSIONS JIGSAW

Working in groups of four or five, students will complete a jigsaw activity where they become an expert in one of the following topics: adding like terms, subtracting like terms, factoring, and expanding. They will begin in their home base group. Each student will select one of the topics in which to become an expert. They will then move to their expert group where they will work collaboratively to learn the material. After they feel comfortable with their material, they will return to their home group and teach the other members of their group by showing examples and creating problems for the other group members to complete. By the end of the activity, each student should feel comfortable with all the information and be able to simplify expressions. *Note:* This activity is likely to run more than one class period.

MATERIALS

Math texts; math reference books; computers with Internet access; algebra tiles.

PROCEDURE

1. After dividing your students into groups, explain that this group is their home group.

2. Place the following topics on the board: adding like terms, subtracting like terms, factoring, expanding. Instruct students to select a topic in which they will become an expert. Ideally, each group member should select a different topic; however, in groups of five, two students may partner and become experts on one topic.

3. Designate parts of the room where the students learning each topic will go. For example, students adding like terms may meet in a corner of the classroom, students working with factoring may meet at the back table, and so on. Instruct students to move into their expert groups.

4. When students meet in their expert groups, they must use the resources provided to learn the material. Along with math texts and math references, following are some resources you may suggest:

   - Adding and subtracting like terms—students may check the following Web sites:
     http://www.algebrahelp.com/lessons/simplifying/combiningliketerms/index.htm
     http://www.freemathhelp.com/combining-like-terms.html
     http://www.purplemath.com/modules/polydefs2.htm
     Each Web site offers a clear explanation about like terms and provides examples for adding and subtracting to simplify expressions.

   - You may also provide algebra tiles to aid your students in their study of combining like terms. Algebra tiles consist of rectangles that represent variables, small squares that represent whole numbers, and larger squares that represent a variable to the second power. Algebra tiles can help students visualize the terms in an expression. By using algebra tiles, students can clearly see what constitutes a like term.

   - Factoring and expanding—students may check this Web site: http://www.hstutorials .net/dialup/distributiveProp.htm. The site contains practice problems and examples.

5. When students feel that they are experts in their chosen topic, instruct them to return to their home group. They will then take turns explaining in detail their topic to the group. They may use any resources that they found helpful for their explanation. They should also provide example problems and practice problems for their group members to complete. The next expert should not begin explaining his or her topic until all group members have mastered the first topic. By the end of this activity, all group members should have shared their topic and all group members should be confident in all topics.

CLOSURE

Have students write a summary of what they have learned about each topic. Their explanations should include the procedures to follow in order to write equivalent expressions as well as a sample problem and solution.

Expressions and Equations: 7.EE.2

"Use properties of operations to generate equivalent expressions."

> 2. "Understand that rewriting an expression in different forms in a problem context can shed light on the problem and how the quantities in it are related."

BACKGROUND

There are several ways to write different quantities. For example, $\frac{3}{4}$ can be expressed as 75%, 0.75, $\frac{1}{4}$ less than 1, three-quarters, and so on. The way we choose to express numbers or expressions changes the way we view and work with those numbers, making a problem easier or more difficult to solve.

 ### ACTIVITY: REWRITING EXPRESSIONS

Working individually, students will write five expressions on index cards, one per card. Working in groups of three or four, they will present their expressions to the other members of the group who will then write these expressions in different forms.

MATERIALS

$4'' \times 6''$ index cards, five per student.

PROCEDURE

1. Provide your students with examples of expressions written in different forms, such as the following:

- $10\% = 0.1 = \frac{1}{10}$

- $150\% = 1.5 = 1 + \frac{1}{2} = \frac{3}{2}$

- $5^2 = 25 = \frac{125}{5} = 2 \times 12\frac{1}{2} = 2 \times 12.5 = \frac{50}{2}$

- $5y + 3y = 8y = 10y - 2y = 4y \times 2$

- $0.25 = \frac{1}{4} = 25\% = 1 - 0.75$

2. Distribute five index cards to each student. Instruct them to write an expression on each one.

3. After students have completed writing their expressions, divide the class into groups. Each group member should present her expressions to the group, one at a time. Group members will then write as many different forms of the expression as they can. After the first student is finished presenting all of her expressions, another group member presents his expressions. This process continues until all members of the group have presented their expressions.

4. Encourage your students to verify the accuracy of the different forms of expressions in their group.

5. Because expressions can be written in many different forms, you should monitor the work of the groups closely for accuracy.

CLOSURE

Ask students to volunteer some of the expressions their group generated, particularly any that were surprising or out of the ordinary. Ask them if some forms of numbers are easier to use than others. Students should provide examples to support their claims.

Expressions and Equations: 7.EE.3

"Solve real-life and mathematical problems using numerical and algebraic expressions and equations."

> 3. "Solve multi-step real-life and mathematical problems posed with positive and negative rational numbers in any form (whole numbers, fractions, and decimals), using tools strategically. Apply properties of operations to calculate with numbers in any form; convert between forms as appropriate; and assess the reasonableness of answers using mental computation and estimation strategies."

BACKGROUND

Estimation is a crucial skill in any math class. It allows students to check the reasonableness of an answer to a mathematical problem and helps them to decide if their answer makes sense.

 ## ACTIVITY: ESTIMATION GAME

Working in groups of four or five, students will participate in a game where they must estimate the answer to a mathematical problem. They must also compute the exact answer and compare this to their estimate.

MATERIALS

Clock or stopwatch.

PREPARATION

Create a list of estimation problems. Obtain a clock or stopwatch that shows seconds.

PROCEDURE

1. Explain that an estimate is an approximate amount. It can be used for checking answers as well as making predictions. Certainly estimation is a part of math curriculums, but it is also used in real-life situations as well.

2. Divide your students into groups. Explain the rules of the game: You will present a math problem. The groups will have thirty seconds to decide on an estimate. (They should not compute the exact answer at this time.) A designated member of the group writes the estimate on a sheet of paper and places it face down. After thirty seconds, ask that a member of each group hold up the paper with the group's estimate. Record

each group's estimate on the board. Now ask the members of each group to work the problem out and compare their estimate to the exact answer. The group whose estimate was closest to the correct answer receives a point. If more than one group has the same closest estimate, each of those groups receives a point. You may present your students with several problems as the game goes on.

3. Following are sample problems that you may choose to use for this game:

   • A dinner bill came to $42. The sales tax was 7%. What is the total cost? ($44.94)

   • The average starting salary for college graduates in 2011 was $51,000. This was a raise of $2,000 from the previous year's starting salary. What was the percent of increase? (4% which is rounded to the nearest percent)

   • Sarah wants to paint her new office. The office's four walls measure 12 feet by 10 feet. One gallon of paint covers 75 square feet. How many gallons of paint will she need to paint her office? (6.4 gallons)

   • John is hanging a picture frame that is 11 inches wide on a wall that measures 6 feet long. He wants to place this picture directly in the middle of the wall. How far away from each side of the wall should John position the picture? (30.5 inches)

   • Movie tickets for a matinee cost $4 per person. The cost of tickets for a movie after 5 pm increases to $10 a person. What is the percent of increase? (150%)

   • If $x = -5$, what is the value of $\dfrac{x}{8} \cdot \dfrac{4}{x}$? $\left(\dfrac{1}{2}\right)$

   • Carly is wrapping a birthday present that will fit in a box (rectangular prism) that measures 1 foot by 9 inches by 4 inches. What is the minimum amount of wrapping paper she will need to wrap the box? ($2.\overline{6}$ square feet)

4. After the game, tally the scores and announce a winner.

CLOSURE

Have your students write a reflection about the advantages and disadvantages of using estimation as a strategy for solving problems. Discuss their answers as a class.

Expressions and Equations: 7.EE.4

"Solve real-life and mathematical problems using numerical and algebraic expressions and equations."

4. "Use variables to represent quantities in a real-world or mathematical problem, and construct simple equations and inequalities to solve problems by reasoning about the quantities.

   a. "Solve word problems leading to equations in the form of $px + q = r$ and $p(x + q) = r$, where p, q, and r are specific rational numbers. Solve equations of these forms fluently. Compare an algebraic solution to an arithmetic solution, identifying the sequence of the operations used in each approach.

   b. "Solve word problems leading to inequalities of the form $px + q > r$ or $px + q < r$, where p, q, and r are specific rational numbers. Graph the solution set of the inequality and interpret it in the context of the problem."

BACKGROUND

To solve an equation, students must isolate the variable. They should simplify each side of the equation, if necessary, then add the same number to or subtract the same number from both sides of the equation. They should then multiply or divide both sides of the equation by the same non-zero number.

To solve an inequality, students should follow the same steps as for solving equations. However, if they are multiplying or dividing both sides of the inequality by a negative number, they must change the direction of the inequality symbol. To graph the solution on a number line, an open circle must be placed on the number line and the number line must be shaded in the direction of the numbers that satisfy the inequality.

For example: $2x + 5 < 4$. The solution is $x < -\dfrac{1}{2}$.

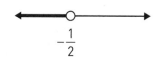

$-\dfrac{1}{2}$

ACTIVITY 1: WRITING EQUATIONS

Working in pairs or groups of three, students will write word problems that can be solved using equations. They will then exchange their problems for the problems of another pair or group of students and solve each other's problems.

Two 4″ × 6″ index cards for each pair or group of students.

PROCEDURE

1. Distribute the index cards to each pair or group of students. Explain that students are to write a word problem that can be solved using an equation of the form $px + q = r$ on one card. On the second index card they are to write a word problem that can be solved using an equation of the form $p(x + q) = r$. They should write the answers to their problems on a separate sheet of paper, which they will retain for later.

2. Offer these examples of word problems that students may use as a guide for writing problems of their own:

   • James worked twelve hours last week at the local ice cream shop. Because of his hard and steady work over the last six months, James's boss gave him a $25 bonus. James received $124 from his boss for last week, which included his bonus. What was James's hourly wage? [Ask your students to write an equation that would help them solve this problem. $px + q = r$ where p represents the number of hours James worked, x represents his hourly wage, q represents the amount of his bonus, and r represents the total amount of money he received. Ask your students to write an equation and solve for x. $\$12x + \$25 = \$124$. $x = \$8.25$.]

   • Sophia works at a local pharmacy. The pharmacy is open on holidays. When Sophia works on a holiday, her boss pays her an additional $3.00 per hour. If Sophia works four hours on a holiday, and she earns $46.40 for that day, what is her normal hourly wage? [Ask your students to write an equation that would help them solve this problem. $p(x + q) = r$, where p represents the number of hours Sophia worked, x represents her normal hourly wage, q represents her additional pay per hour, and r represents her total pay for the day. Ask your students to write an equation and solve for x. $4(x + \$3) = \$46.40$. $x = \$8.60$.]

3. After students have completed writing their problems, instruct them to exchange index cards with another pair or group of students.

4. Students are to write an equation for the word problems they received, then solve the problems.

5. After students have completed the problems on both index cards, they should conference with the students they received their index cards from. They should discuss the problems, equations, and solutions, and verify the correctness of the answers (referring to the original answers of the students who wrote the problems).

CLOSURE

Ask some students to share their word problems and equations with the class. They may come to the front of the room and write their equations and solutions on the board. Review the sequence of operations that were necessary to solve the problems.

ACTIVITY 2: SOLVING INEQUALITIES

This is a two-day activity. Working in pairs or groups of three, students will write word problems that can be solved by using inequalities. After students have completed their problems, the teacher will project the problems on the board while students solve and graph the inequalities on whiteboards.

MATERIALS

One transparency for each pair or group of students; non-permanent markers; small whiteboards for students to use (one per individual or one per pair or group of students); dry-erase markers; erasers or tissues.

PROCEDURE

1. On day one, distribute the transparencies, non-permanent markers, and erasers or tissues to your students. Explain that they are to write a word problem that can be solved using an inequality of the form $px + q > r$ or $px + q < r$ on their transparency. They should write neatly and place their names and the answers to their problems in a corner of the transparency.

2. Offer this example of a word problem that students may use as a guide for writing problems of their own:

   - Inez wants to buy a new cell phone. (Her parents will pay her monthly service bill, but she must buy the phone with her own money.) The new phone costs slightly more than $300. Inez already has saved $176.00. She babysits after school for a neighbor three days each week and earns $36.00 per week. At this rate, how many weeks must she babysit to have enough money to buy the phone? [Ask your students to write an inequality that could be used to solve this problem, $px + q > r$, where p represents Inez's earnings each week from babysitting, x represents the number of weeks she works, q represents the amount of money she already has saved, and r represents the cost of the phone. Ask your students to write and solve an inequality for x. $36x + \$176 > \$300$. $x > 3\frac{4}{9}$ weeks, but because she is paid weekly, she must work 4 weeks.]

3. After students have completed their problems, collect the transparencies. Check that their equations and answers are correct. If necessary, write suggestions on the transparency as to how a problem might be corrected or improved.

4. On the next day, hand back the transparencies and distribute all the other materials. Allow a few minutes for groups to make any revisions, if necessary. Re-collect the transparencies and project and read each word problem, one at a time, to your students. Be sure to block the answer. Working individually (or in pairs or small groups), students are to solve the problem, then graph the solution on a number line on their whiteboards. (Instead of the whiteboards, you may substitute clear page protectors. Place a white sheet of paper behind the page protector for a background.)

5. After each problem is solved, have your students hold up their whiteboards with the solution and graph of the inequality.

6. Continue this procedure with additional word problems.

CLOSURE

Pose questions such as the following: How does an inequality differ from an equation? How do you solve inequalities? What are some important things to remember when graphing inequalities?

Geometry: 7.G.1

"Draw, construct, and describe geometrical figures and describe the relationships between them."

> 1. "Solve problems involving scale drawings of geometric figures, including computing actual lengths and areas from a scale drawing and reproducing a scale drawing at a different scale."

BACKGROUND

A scale drawing represents the actual size of an object or figure that has been reduced or enlarged by a specific amount, called a scale factor. The purpose of a scale drawing is to create a drawing that has the same proportions as the original it represents. The ratio used in a scale drawing is the drawing's length to the actual length of the object.

To create a scale drawing, you must know the actual dimensions of the object and the size of the paper. Then you must choose an appropriate scale so that the drawing will fit on your paper. For example, if the dimensions of your classroom are 12 feet by 15 feet, you may choose the scale 1 inch to 1.5 feet or 1 inch to 18 inches. This means that every inch on the drawing represents 18 inches of the classroom. Using this scale, the length of the classroom on your scale drawing is 8 inches and the width is 10 inches.

You can also use a scale to determine actual lengths. For example, if the scale is 1 to x, multiply the length of each dimension on the scale by x to find the actual dimensions. Using the previous example, the scale drawing is 8 inches by 10 inches. The scale is 1:18 inches. To find the length of the room, multiply 18 by 8, which equals 144 inches, which in turn equals 12 feet. To find the width of the room, multiply 18 by 10, which equals 180 inches or 15 feet.

 ## ACTIVITY: SCALING YOUR CLASSROOM

Working individually as well as in pairs or groups of three, students will use a scale drawing to calculate the dimensions of the classroom. They will then use the actual dimensions to create a scale drawing at a different scale.

MATERIALS

Copies of a scale drawing of the classroom; yardsticks or meter sticks; unlined paper.

PREPARATION

Measure your classroom and create a scale drawing (including the scale you used) that you will present to your students. This drawing could be as basic or involved as you would like. You may wish to simply scale the dimensions of the room, or you may wish to include desks, tables, bookshelves, and so on (viewed from the top).

PROCEDURE

1. Distribute copies of your scale drawing to your students. Explain what the scale is and show them its location on your drawing.

2. Instruct your students to work individually and use the scale you provided to find the actual dimensions of the room. (If you have included other items in the scale drawing, you might also ask students to find the areas of the tops of these items.) Using the actual dimensions, they are to then find the area of the room.

3. After students have calculated the actual dimensions, have them work in pairs or groups of three to verify the dimensions they found. Instruct them to use yardsticks or meter sticks to measure the actual dimensions in the room and calculate the area of the room. (Note: Depending on how many students you have in your class, you may prefer to select only a few students to measure the dimensions and objects in the room and report to the class.)

4. Confirm the actual dimensions of the room with your students and allow them to correct their work, if necessary.

5. Once students know the actual dimensions of the classroom, instruct them to select a different scale and create a new scale drawing using this new scale. You may wish to have some students select a larger scale while others choose a smaller scale. Depending on the scale students select, you may need to provide larger sheets of paper. (Their drawings might not fit on a standard sheet of paper.)

6. Once their scale drawings are complete, have students work in pairs or groups of three once again to check each other's work, particularly that their scales multiply to the actual dimensions.

CLOSURE

Ask for volunteers to share their scale drawings with the class. Discuss the uses of scale drawings in real-world situations—for example, on maps, diagrams, and models.

Geometry: 7.G.2

"Draw, construct, and describe geometrical figures and describe the relationships between them."

2. "Draw (freehand, with ruler and protractor, and with technology) geometric shapes with given conditions. Focus on constructing triangles from three measures of angles or sides, noticing when the conditions determine a unique triangle, more than one triangle, or no triangle."

BACKGROUND

Given certain conditions students can easily classify and draw geometric shapes. For example, the sum of the measures of the interior angles of a triangle equals 180°. The sum of any two sides of a triangle is always longer than the other side.

Triangles can be classified by the measures of their angles and lengths of their sides. When triangles are classified according to the types of angles, triangles may be acute, obtuse, or right. In an acute triangle, all angles are less than 90°. In an obtuse triangle, one angle is larger than 90°. In a right triangle, the measure of one angle is 90°. When triangles are classified according to the lengths of their sides, they may be scalene, equilateral, or isosceles. The sides of scalene triangles have three different lengths. The lengths of all three sides of equilateral triangles are congruent. Isosceles triangles have two congruent sides.

 ## ACTIVITY: CREATING TRIANGLES

Students will work in groups of three or four. They will draw triangles according to given conditions and classify each triangle based on the conditions. They will draw triangles in free hand, with rulers and protractors, and with computer software.

MATERIALS

White construction paper; rulers; protractors; computers with software such as Geometer's Sketchpad that enables users to draw accurate geometric figures; reproducible, "Investigating Triangles."

Set up three different stations, one at which students will draw triangles free hand, another at which students will draw triangles with rulers and protractors, and a third at which they will draw triangles with computers and appropriate software. (Note: Depending on the size of your class, you may want to set up two of each station.) Make enough copies of the reproducible so that each station has a list of the triangles students are to draw.

PROCEDURE

1. Review the different types of triangles with your students.

2. Explain that students will be working in groups of three or four and will travel to three different stations where they will draw various triangles, given specific conditions.

3. Instruct the groups to go to one station at a time and work with one triangle at a time. Students are to draw each of the triangles at every station. Caution your students that not all of the triangles can be drawn according to the given conditions.

CLOSURE

Ask your students questions such as the following: Which triangle or triangles could not be drawn with the given conditions? Explain why not. Which triangles were easiest to draw? Why? Which were difficult to draw? Why? Which method—free hand, ruler and protractor, or computer software—was most accurate for drawing triangles? Designate a section of a bulletin board in the classroom for students to hang their drawings of triangles.

INVESTIGATING TRIANGLES

- An acute triangle with one angle that measures 38°.

- A scalene triangle with one side that measures $4\frac{1}{2}$ inches.

- A right triangle with one angle that measures 60° and another angle that measures 35°.

- An obtuse triangle with one angle that measures 70°.

- An equilateral triangle with one side that measures 3 inches.

- An isosceles triangle with one side that measures 5 inches.

- A right triangle with two sides that each measure $3\frac{1}{4}$ inches.

- An acute triangle with one angle that measures 38°.

- A scalene triangle with one side that measures $4\frac{1}{2}$ inches.

- A right triangle with one angle that measures 60° and another angle that measures 35°.

- An obtuse triangle with one angle that measures 70°.

- An equilateral triangle with one side that measures 3 inches.

- An isosceles triangle with one side that measures 5 inches.

- A right triangle with two sides that each measure $3\frac{1}{4}$ inches.

- An acute triangle with one angle that measures 38°.

- A scalene triangle with one side that measures $4\frac{1}{2}$ inches.

- A right triangle with one angle that measures 60° and another angle that measures 35°.

- An obtuse triangle with one angle that measures 70°.

- An equilateral triangle with one side that measures 3 inches.

- An isosceles triangle with one side that measures 5 inches.

- A right triangle with two sides that each measure $3\frac{1}{4}$ inches.

Geometry: 7.G.3

"Draw, construct, and describe geometrical figures and describe the relationships between them."

> 3. "Describe the two-dimensional figures that result from slicing three-dimensional figures, as in plane sections of right rectangular prisms and right rectangular pyramids."

BACKGROUND

When three dimensional figures are sliced at different angles, various two-dimensional figures are formed by the cross section. The resulting figures vary according to where the plane intersects the figures. Following is a list of some three-dimensional figures and some two-dimensional figures that result from slicing the three-dimensional figures:

- *Cone* — circle, triangle, ellipse
- *Cylinder* — circle, rectangle
- *Rectangular prism* — rectangle
- *Right rectangular pyramid* — triangle, rectangle
- *Cube* — square, triangle, pentagon, hexagon, rectangle, parallelogram, trapezoid

 ACTIVITY 1: A VIRTUAL CUBE

Working individually or in pairs, students will virtually slice a cube and record the figure that is formed by a cross section of the slice.

MATERIALS

Computers with Internet access.

PROCEDURE

1. Instruct students to go to the Web site http://nlvm.usu.edu/en/nav/vlibrary.html. They should click on "Geometry, Grades 6–8," then scroll down and click on "Platonic Solids—Slicing."

2. Instruct students to click on "cube" (if the cube is not already on the screen). A cube will appear on the left. Explain that the cube will be sliced by a plane that is parallel to the computer screen. The outline of the figure formed by the cross-section slice is displayed on the right.

3. Instruct your students to rotate the cube by clicking and dragging the cursor anywhere on the cube. Students will see the outline of the figure formed by the cross section. They should sketch and record the figure on paper.

4. Allow your students to explore the figures of the cross sections by rotating the cube. They should continue sketching and recording the different figures they discover.

CLOSURE

Compile a class list of the various figures of cross sections.

ACTIVITY 2: SLICING FIGURES

Students will create three-dimensional figures out of modeling clay. They will then use a large, flattened paper clip to slice the model and record the resulting two-dimensional figure.

MATERIALS

Modeling clay (4–6 ounces per student); large paper clips; paper towels (for cleanup).

PROCEDURE

1. Distribute clay, a large paper clip, and a paper towel to each student. (Have extra materials, especially paper towels, on hand.)

2. Explain that students will use the clay and paper clips to create three- and two-dimensional figures. They are to create the following figures, one at a time: a cone, cylinder, rectangular prism, right rectangular pyramid, and cube. If necessary, provide examples of these figures.

3. Once students are finished making the first figure, they will use their paper clip to slice their three-dimensional figure at any angle. Then they will record the resulting two-dimensional figure formed by the cross section by drawing the two-dimensional figure and naming it. Students should reconstruct their three-dimensional figure and slice it again. Challenge your students to create as many two-dimensional figures for each three-dimensional figure as possible. Students should repeat this process for the remaining figures.

4. At the conclusion of the activity, remind students to clean up neatly.

CLOSURE

Have a class discussion about students' findings. You may wish to have several students come to the front of the room to demonstrate how they sliced a figure and show the figure that resulted from it. Ask students how they can predict what two-dimensional figure will result from slicing a three-dimensional figure. Did they notice any patterns? If yes, they should describe them.

Geometry: 7.G.4

"Solve real-life and mathematical problems involving angle measure, area, surface area, and volume."

4. "Know the formulas for the area and circumference of a circle and use them to solve problems; give an informal derivation of the relationship between the circumference and area of a circle."

BACKGROUND

The circumference of a circle is the distance around a circle. It is represented by the letter C. The circumference can be found by using the formula $C = \pi d$, where d is the diameter of the circle, or by $C = 2\pi r$, where r is the radius.

The area of a circle is the number of square units inside the circle. It is found by using the formula $A = \pi r^2$, where A is the area and r is the radius of the circle.

The relationship between the circumference and area of a circle is given by the formulas $C = 2\sqrt{A\pi}$ or $A = \dfrac{C^2}{4\pi}$.

 ACTIVITY 1: CIRCLE SCAVENGER HUNT

Students will find circles in and around their home. They must identify each circle, measure its diameter or radius, and calculate its circumference and area. The next day they will verify their answers by using a virtual interactive circle.

MATERIALS

Rulers; calculators; computers with Internet access.

PROCEDURE

1. Review the definitions and formulas for finding the circumference and area of circles. Emphasize what each variable represents.

2. Challenge your students to find at least five circles in or around their homes. These circles could be on common objects such as erasers, magnets, or they could be represented by bottle tops, plates, or pizza pies.

3. For each circle a student finds, she must measure its radius or diameter and calculate its circumference and area. Students should record all their measurements, as well as the objects they measure.

4. When students return to class, instruct them to go to the Web site www.mathwarehouse .com/geometry/circle/interactive-circumference.php where they will check the circumferences of the circles they found. Explain that by clicking and dragging point B on the circle, they can create a circle that has the same radius or diameter as a circle they found. They can check the circumference with the Web site's generated responses.

5. After they have checked the circumferences of their circles, instruct them to go to the Web site www.mathwarehouse.com/geometry/circle/area-of-circle.php where they can check the areas of the circles they found, following the same procedure as in step 4.

CLOSURE

Ask for volunteers to share examples of the various circles they found. Students will no doubt be surprised at how common this geometric shape is. Ask who found the largest and smallest circles. (Instruct your students to retain their work, which will be used in Activity 2.)

ACTIVITY 2: WHAT'S THE RELATIONSHIP?

Students will work in pairs or groups of three to find the relationship between the area and circumference of the circles they found in Activity 1. They will look for patterns and derive the formula algebraically.

MATERIALS

The circumferences and areas of circles students found in Activity 1.

PROCEDURE

1. Explain that the circumference and area of a circle are related. Students will use the circumferences and areas that they found in Activity 1 to find this relationship.

2. Allow time for your students to look for patterns and try to determine the relationship, $C = 2\sqrt{A\pi}$ or $A = \dfrac{C^2}{4\pi}$. Because the relationships are nonlinear, students may have difficulty discovering these formulas. You may provide some prompts such as:

   • How does the circumference compare to the square root of the area?
     (Answer: $C = 2\sqrt{\pi} \cdot \sqrt{A}$ or about $3.5\sqrt{A}$)

   • How does the area compare to the circumference squared? (Answer: $\dfrac{1}{4\pi} \cdot C^2$ or about $0.08C^2$)

3. Ask your students to derive the relationships, following the steps below:

- Express each formula in terms of r. $\left(r = \dfrac{C}{2\pi} ; r = \sqrt{\dfrac{A}{\pi}} \right)$

- Substitute each expression for r. $\left(\dfrac{C}{2\pi} = \sqrt{\dfrac{A}{\pi}} \right)$

- Square each side. $\left(\dfrac{C^2}{4\pi^2} = \dfrac{A}{\pi} \right)$

- Solve for C or solve for A. $\left(C = 2\sqrt{A\pi} ; A = \dfrac{C^2}{4\pi} \right)$

CLOSURE

Ask your students to explain whether deriving the formula for relating the circumference and area makes the relationship clearer. What, if any, is the benefit of being able to derive a formula from patterns and relationships? (Answers may vary, but students should realize that being able to derive formulas provides mathematical insight and deeper understanding of mathematical relationships.)

Geometry: 7.G.5

"Solve real-life and mathematical problems involving angle measure, area, surface area, and volume."

> 5. "Use facts about supplementary, complementary, vertical, and adjacent angles in a multi-step problem to write and solve simple equations for an unknown angle in a figure."

BACKGROUND

Some pairs of angles have their own name and special features.

- *Supplementary angles* are two angles whose measures add up to 180°.
- *Complementary angles* are two angles whose measures add up to 90°.
- *Adjacent angles* are two angles that have the same vertex and a common side.
- *Vertical angles* are two nonadjacent angles that are formed by two intersecting lines.

ACTIVITY: WHAT'S THE ANGLE?

Students will work in pairs or groups of three to identify which equations and angle measures can be matched to a sketch of a figure.

MATERIALS

Scissors; glue sticks; sheets of unlined paper; reproducible, "Angle Measures and Equations."

PROCEDURE

1. Explain to your students that, when given a sketch or figure, they may write and solve equations to find the measures of missing angles.

2. Distribute the materials and a copy of the reproducible to each group. Explain that the reproducible contains four sketches and 24 equation cards. Students are to cut out each equation card. They should not cut out the sketches.

3. Explain that the sketches are not drawn to scale and that protractors cannot be used to find the measure of an angle. Students can assume that all lines are straight lines and all right angles are represented by a small box in the vertex. Some equation cards are identical, because the same equation can match two sketches.

4. Depending on your students, you may wish to review pairs of angles and their properties, as noted in the Background.

5. Instruct your students to glue the sketches across the top of another sheet of paper. They are now to match each equation card with a sketch, then glue the matching cards beneath each sketch.

CLOSURE

Review the equations students have matched with each sketch. Ask students if they can write any additional equations that were not included on the cards.

ANSWERS

Sketch 1: $y + 90° = 180°$; $x + 90° = 180°$; $180° - 90° = x$; $x = y$; $x = 90°$; $y = 90°$

Sketch 2: $x + y = 180°$; $50° + x = 180°$; $180° - 50° = x$; $x = 130°$; $y = 50°$

Sketch 3: $x + y + 50° = 180°$; $y + 50° = 140°$; $x + 50° = 90°$; $90° - x = 50°$; $x = 40°$; $y = 90°$

Sketch 4: $x + 40° = 130°$; $x + y + 50° = 180°$; $x + 40° + 50° = 180°$; $y + 50° = 90°$; $50° + x = 140°$; $x = 90°$; $y = 40°$

ANGLE MEASURES AND EQUATIONS

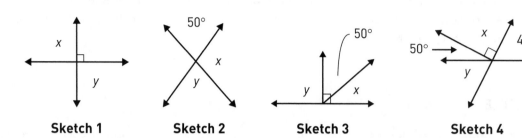

Sketch 1	Sketch 2	Sketch 3	Sketch 4

$50° + x = 180°$	$x + y + 50° = 180°$	$y = 90°$	$y + 50° = 140°$
$X = 90°$	$y = 50°$	$y = 40°$	$x = y$
$180° - 90° = x$	$y + 90° = 180°$	$x = 40°$	$x + 40° + 50° = 180°$
$X + 50° = 90°$	$x + y = 180°$	$50° + x = 140°$	$x + 90° = 180°$
$90° - x = 50°$	$x + 40° = 130°$	$x = 130°$	$x + y + 50° = 180°$
$X = 90°$	$y = 90°$	$y + 50° = 90°$	$180° - 50° = x$

Geometry: 7.G.6

"Solve real-life and mathematical problems involving angle measure, area, surface area, and volume."

> 6. "Solve real-world and mathematical problems involving area, volume, and surface area of two- and three-dimensional objects composed of triangles, quadrilaterals, polygons, cubes, and right prisms."

BACKGROUND

Some formulas for finding the areas of polygons are listed below:

- Triangle: $A = \frac{1}{2}bh$

- Square: $A = s^2$

- Rectangle: $A = l \times w$

- Trapezoid: $A = \frac{1}{2}h(b_1 + b_2)$

- Parallelogram: $A = bh$

To find the areas of irregular figures (figures composed of one or more of the figures listed above), students should break up the figures into polygons that they can find the areas of and then add the areas to find the total area.

To find the volume of prisms, students should find the area of the base and multiply it by the height, $V = Bh$, where B stands for the area of the base, and h stands for the height of the prism.

To find the surface area of three-dimensional figures, students should find the sum of the areas of all the faces and the base or bases.

 ## ACTIVITY: LET'S BUILD IT

Working in pairs or groups of three, students will create one two-dimensional figure using Unifix® cubes. They will then build a three-dimensional figure by stacking additional rows of cubes. They will find the volume and surface area of the figure.

MATERIALS

Unifix® (or other interlocking cubes), about 30 for each pair or group of students.

PROCEDURE

1. Explain that a two-dimensional figure has length and width, but no height. A three-dimensional figure has length, width, and height.

2. Explain that every three-dimensional object composed of polygons has at least one base that has two dimensions.

3. Instruct your students to use 6 cubes to form a prism whose base is a 3 × 2 rectangle. Viewing the rectangle from above, students will see a rectangle whose area is 6 square units. This is the area of the base. By taking into account the rectangle's height, the three dimensions of the object become apparent. The volume of the object (a rectangular prism) is found by multiplying the area of the base by the height. 6 square units × 1 unit = 6 cubic units.

4. Explain that to find the surface area of a rectangular prism, students must add the area of all the faces, including the base or bases. Following is an example:

Area of the top:	2 × 3 = 6
Area of the bottom:	2 × 3 = 6
Area of one face:	2 × 1 = 2
Area of the opposite face:	2 × 1 = 2
Area of another face:	3 × 1 = 3
Area of the opposite face:	3 × 1 = 3
Total:	22 square units

5. Once students understand the procedure, encourage them to build a three-dimensional figure of their own with Unifix® cubes. They are to find the area of the base, volume of the figure, and its surface area, recording their work and answers.

6. After students have built their figures and made their calculations, instruct them to exchange their work with another pair or group of students and check each other's answers. They should correct any mistakes.

7. While working with another pair or group, students should combine their figures. Note that they should not rearrange their figures, but simply combine them to form an irregular figure. Working together, they are to find the surface area and volume of their combined figure.

CLOSURE

Discuss students' results. Ask your students what they noticed about the volume and surface area of the combined figures compared with their original figures. (Their answers will vary, depending on their figures.) To conclude the activity, ask students to explain how to find the surface area and volume of a prism that has a triangular base. (Answers will vary, but students should note that to find the surface area of the prism they must find the areas of the two triangular bases and add them to the areas of the three faces. To find the volume, students should find the area of the base and multiply it by the height of the figure.)

Statistics and Probability: 7.SP.1

"Use random sampling to draw inferences about a population."

1. "Understand that statistics can be used to gain information about a population by examining a sample of the population; generalizations about a population from a sample are valid only if the sample is representative of that population. Understand that random sampling tends to produce representative samples and support valid inferences."

BACKGROUND

Statisticians analyze data to gain knowledge and make predictions about a population. But the typical population is so large that it is nearly impossible to examine it as a whole. Statisticians therefore use a sample—a subset of the population—because the sample provides a manageable amount of data that represents the population. The best samples are random samples. In a random sample, each item is chosen entirely by chance, and each item has the same probability of being selected for the sample.

 ACTIVITY: EXAMINING SAMPLES

Students will participate in a survey. Working in groups of three or four, they will use the results of the survey to analyze a sample of a population and use that information to make inferences about the population.

MATERIALS

Copies of the results of the survey; computer and printer.

PREPARATION

A poll of students and a copy of the results of the survey for each student.

PROCEDURE

1. Select a question to ask your students a few days before you begin this activity. Example questions include:

* About how many hours of TV do you watch each day?

* About how much time do you spend doing homework each day?

* About how many hours did you play or watch a sport this week?

2. Use this question to poll your students. Instruct them to hand in a sheet of paper with their name and answer on it.

3. Record the results on a spreadsheet based on the order you receive students' papers. Include the results for all your classes because this activity will work best with a large data set. Instead of writing each student's name, write Student 1, Student 2, and so on. Also, separate the results for the boys and girls by placing the boys' responses in one column and the girls' responses in another.

4. To begin the activity, hand out copies of the results of the poll to your students. Explain that you have recorded the results of the poll and that they will analyze the results by looking at a sample of the data. Review the definitions of a sample and population.

5. Assign each group a different sample of the data to analyze. For example:

   - Group 1 analyzes the boys' responses.
   - Group 2 analyzes the girls' responses.
   - Group 3 analyzes the first quarter of the total data.
   - Group 4 analyzes the second quarter of the total data.
   - Group 5 analyzes the third quarter of the total data.
   - Group 6 analyzes the fourth quarter of the total data.

6. As they are working in their groups, students should analyze the data of their sample and draw inferences about the data.

7. After students have their results, each group should report to the class what sample they analyzed and what inferences they drew. Some groups may have similar answers; other groups may have very different answers.

CLOSURE

Instruct your students to write an explanation of why it is necessary to use random sampling when analyzing a large population.

Statistics and Probability: 7.SP.2

"Use random sampling to draw inferences about a population."

> 2. "Use data from a random sample to draw inferences about a population with an unknown characteristic of interest. Generate multiple samples (or simulated samples) of the same size to gauge the variation in estimates or predictions."

BACKGROUND

The entire group of objects or individuals from which a statistical sample is taken is called a population. A sample is a group of objects or individuals that are chosen from the population to make predictions about the population. A random sample is a group that is chosen by chance.

 ## ACTIVITY: HOW MANY CUBES?

Working in pairs or groups of three, students will randomly select different-colored cubes from a large container and make inferences about the number of cubes in the container.

MATERIALS

About 150 different-colored cubes such as Snap Cubes®, math linking cubes, or centimeter cubes; an opaque container (or a clear container wrapped in paper) large enough to hold all of the cubes.

PREPARATION

If necessary, wrap the container with paper to prevent students from seeing the contents. Count and record the number of colored cubes you placed in the container.

PROCEDURE

1. Tell your students that different-colored cubes are in the container and that you know how many of each color there are. Shake the container to mix the contents.

2. Explain that each pair or group of students is to select 15 cubes from the container (without looking inside), count, and record the numbers and colors of the cubes they selected. For example, they might have picked 4 red, 6 blue, and 5 white. They are to then return the cubes to the container.

3. Shake the container again and have another pair or group of students select 15 cubes (without looking inside), count, and record the numbers and colors of the cubes. This process continues until all students have selected and recorded the numbers and colors of the cubes they selected.

4. Instruct your students to analyze their data and make predictions about the numbers of different-colored cubes in the container. For example, if a pair of students selected 4 white cubes, 7 red cubes, and 3 blue cubes, they may conclude that half of the cubes in the container are red. They may also conclude that there are only three different-colored cubes in the container.

5. Reveal the total number of cubes and the total numbers of different-colored cubes in the container. Ask your students how close their predictions came to the actual numbers.

CLOSURE

Discuss how and why the results of your students may have varied. Instruct them to write a summary describing this experiment in terms of population and random samples.

Statistics and Probability: 7.SP.3

"Draw informal comparative inferences about two populations."

3. "Informally assess the degree of visual overlap of two numerical data distributions with similar variabilities, measuring the difference between the centers by expressing it as a multiple of a measure of variability."

BACKGROUND

Data can be used for making a comparison between two populations in a variety of ways. One is by comparing a measure of center (mean, median, and mode) of each group. When measures of center are coupled with a measure of variability such as the mean absolute deviation, which measures how far on average the data points vary from the mean, a better analysis of the data is possible. A dot plot provides a visual display of the data.

 ## ACTIVITY: HOW WELL DID THEY DO?

Working in groups of three or four, students will determine which class did better on a data analysis test. They will use a measure of center and the mean absolute variation to describe the data and draw two dot plots to represent the data.

MATERIALS

Rulers; reproducible, "Analyzing Test Scores."

PROCEDURE

1. Explain to your students that Mrs. Mean teaches two small classes. She recently gave a test to her first-period and third-period classes. The results are included on the reproducible.

2. Distribute copies of the reproducible and explain that students are to analyze each group of scores according to the directions. Depending on your students, review how to find the measures of center (mean, median, and mode) and the mean absolute deviation. Students must also draw a dot plot for each class's scores.

Have your students report their explanations and share their dot plots. Student explanations may vary. Following are the dot plots for the test scores of Mrs. Mean's Period 1 and Period 3 classes:

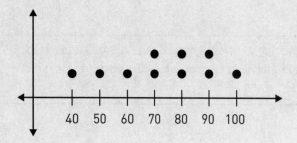
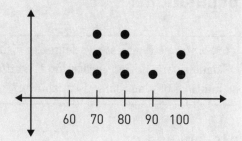

ANALYZING TEST SCORES

> The test scores for Mrs. Mean's two classes are listed below.
>
> **Period 1:** 90, 70, 60, 80, 100, 80, 70, 40, 50, 90
>
> **Period 3:** 100, 90, 70, 60, 80, 100, 80, 80, 70, 70

Your task is to write an explanation comparing the scores of each class. Then display the data on a dot plot.

Follow the guidelines below for writing your explanations:

1. Select one of the measures of center (mean, median, or mode) to compare the data of each class.

2. Find the mean absolute deviation to indicate whether the scores are close together or far apart. A small mean absolute deviation indicates that the data are close together. A large mean absolute deviation indicates that the data are spread out. Compare the mean absolute deviations of each class.

Follow the guidelines below for constructing a dot plot:

1. Title the graph.

2. Place the scores on a vertical number line, using tick marks to represent intervals of ten units.

3. Indicate the frequency of each score by placing a dot above the score.

Statistics and Probability: 7.SP.4

"Draw informal comparative inferences about two populations."

> 4. "Use measures of center and measures of variability for numerical data from random samples to draw informal comparative inferences about two populations."

BACKGROUND

By looking at a small amount of data that has been randomly acquired, students can use the measures of center and measures of variability to make a generalization about the larger group. The data used in this activity was obtained by comparing the labels on randomly selected name-brand cereals with comparable store-brand cereals.

 ACTIVITY: WHAT CAN WE SAY?

Working in pairs or groups of three, students will analyze the nutritional value of a variety of store-brand cereals and name-brand cereals. They will draw comparative inferences using the measures of center and measures of variability to support their claims.

MATERIALS

Reproducible, "Nutritional Values of Randomly Selected Cereals."

PROCEDURE

1. Explain that by selecting a random sample of a population, inferences can be made about the larger group. Polls are based on this method.

2. Explain that the purpose of this activity is to compare store-brand cereals with name-brand cereals. Note that the data for this activity was obtained by comparing the labels of randomly selected store-brand and name-brand cereals. If necessary, explain the difference between the two types of cereals.

3. Distribute a copy of the reproducible to each pair or group of students. Fourteen cereals are listed. The data includes serving size (comparable cereals have the same serving size), calories, total fat, carbohydrates, fiber, and sugar. Note that all of the name-brand cereals are listed first and are followed by the store-brand cereals. The first name-brand cereal is comparable with the first store-brand cereal. For example, Cheerios is a name-brand cereal. The comparable store-brand cereal is Toasted Oats. The second name-brand cereal is comparable with the second store-brand cereal, and so on.

4. Explain that students will select two categories from the data, analyze the measures of center and measures of variability—for all of the name-brand cereals, then for all of the store-brand cereals—and make an inference about how the store-brand cereals compare with the name-brand cereals. For example, if students decide to select "calories," they should find the average of the calories in the store-brand cereal and compare it with the average of calories in the name-brand cereal. (Or they may find the median or the mode of the calories and make a comparison between the name-brand cereal and the store-brand cereal.) To make a determination of how far the data is spread, they could find the range of calories for each group or find the mean absolute deviation. After finding these values, students are to make a conjecture about how the store-brand cereal compares with the name-brand cereal and explain why they made this conjecture.

CLOSURE

Have your students report their results to the class. Ask if they can conclude that their conjectures can apply to all store-brand cereals.

NUTRITIONAL VALUES OF RANDOMLY SELECTED CEREALS

Name-Brand Cereals	Serving Size (Cups)	Calories	Total Fat (grams)	Carbs (grams)	Fiber (grams)	Sugar (grams)
Cheerios® (General Mills)	1	100	2	20	3	1
Corn Flakes® (Kellogg's)	1	100	0	24	1	2
Fruit Loops® (Kellogg's)	1	120	1	26	1	13
Rice Krispies® (Kellogg's)	$1\frac{1}{4}$	120	0	29	0	3
Coco Puffs® (Kellogg's)	$\frac{3}{4}$	110	$1\frac{1}{2}$	23	1	12
Lucky Charms® (General Mills)	$\frac{3}{4}$	110	1	22	1	11
Cinnamon Toast Crunch® (General Mills)	$\frac{3}{4}$	130	3	25	1	10
Store-Brand Cereals						
Toasted Oats	1	100	$1\frac{1}{2}$	21	3	1
Corn Flakes	1	100	0	24	<1	2
Fruit Floats	1	110	1	25	<1	12
Crispy Rice	$1\frac{1}{4}$	120	0	29	0	3
Coco Crispy Rice	$\frac{3}{4}$	120	1	27	0	14
Marshmallow Dreams	$\frac{3}{4}$	120	1	27	1	13
Apple Cinnamon Toasted Oats	$\frac{3}{4}$	120	$1\frac{1}{2}$	26	1	13

Statistics and Probability: 7.SP.5

"Investigate chance processes and develop, use, and evaluate probability models."

5. "Understand that the probability of a chance event is a number between 0 and 1 that expresses the likelihood of the event occurring. Larger numbers indicate greater likelihood. A probability near 0 indicates an unlikely event, a probability around $\frac{1}{2}$ indicates an event that is neither unlikely nor likely, and a probability near 1 indicates a likely event."

BACKGROUND

Probability is defined as the chance that an event or outcome will occur. To determine the probability of an event, write the ratio of the number of favorable outcomes to the number of possible outcomes. The probability of an event, E, is given by the formula $P(E) = \dfrac{\text{number of favorable outcomes}}{\text{number of possible outcomes}}$. Probabilities can range from 0 (which is impossible) to 1 (which is certain).

ACTIVITY: ON A SCALE OF ZERO TO ONE

Students will work in groups of four or five to determine the probabilities of different situations. Each group will be assigned a situation to investigate. They must put the probabilities in order, from the smallest chance of occurring to the greatest chance. They will then report their findings to the class.

MATERIALS

Three decks of playing cards (with Jokers removed); two number cubes, each numbered 1 through 6 (or one pair of dice); 2-page reproducible, "Probability Investigations."

PREPARATION

Make one copy of the reproducible. Cut out each investigation task by cutting on the dotted lines.

PROCEDURE

1. Assign each group one of the following investigations:

   * Investigation 1: A deck of cards

   * Investigation 2: A deck of cards

- Investigation 3: A deck of cards

- Investigation 4: A number cube

- Investigation 5: A number cube

2. Distribute an investigation and the materials for each task to each group. Explain to your students that they are to find the probabilities posed by the questions. Review the notation for expressing probability, for example, P(diamond) means the probability of selecting a diamond. Remind them that after finding the probabilities, they must place the probabilities in order from least to greatest. They must also be prepared to explain how they determined the probabilities of the events.

3. Instruct your students to complete each investigation and record their results.

4. After all groups have completed their investigation, discuss their results as a class. Record the results on the board. Conduct a class discussion about the various probabilities. Ask your students questions such as the following: Which probabilities have the greatest chance of occurring? Which probabilities have the smallest chance of occurring? Are any probabilities equally likely? Discuss how students determined the order of the probabilities. What does this order represent?

CLOSURE

Instruct your students to develop situations of their own that have a probability of 0, 1, and $\frac{1}{2}$. Ask them to hand in their responses before they leave. As a follow-up, you may discuss these situations the next day.

ANSWERS FOR THE INVESTIGATIONS

Investigation 1:   1) $\frac{1}{4}$ 2) $\frac{5}{13}$ 3) $\frac{1}{52}$ 4) 0 5) $\frac{9}{13}$ 6) $0, \frac{1}{52}, \frac{1}{4}, \frac{5}{13}, \frac{9}{13}$

Investigation 2:   1) $\frac{1}{4}$ 2) $\frac{4}{13}$ 3) $\frac{1}{52}$ 4) 1 5) 0 6) $0, \frac{1}{52}, \frac{1}{4}, \frac{4}{13}, 1$

Investigation 3:   1) 0 2) $\frac{1}{4}$ 3) $\frac{3}{13}$ 4) $\frac{1}{52}$ 5) $\frac{1}{2}$ 6) $0, \frac{1}{52}, \frac{3}{13}, \frac{1}{4}, \frac{1}{2}$

Investigation 4:   1) $\frac{1}{6}$ 2) $\frac{1}{2}$ 3) $\frac{1}{3}$ 4) 0 5) $\frac{5}{6}$ 6) $0, \frac{1}{6}, \frac{1}{3}, \frac{1}{2}, \frac{5}{6}$

Investigation 5:   1) $\frac{1}{6}$ 2) $\frac{1}{2}$ 3) 1 4) $\frac{2}{3}$ 5) 0 6) $0, \frac{1}{6}, \frac{1}{2}, \frac{2}{3}, 1$

PROBABILITY INVESTIGATIONS

- ✂ - - -

INVESTIGATION 1: DECK OF CARDS

Use your playing cards to determine the probability of each event. Note that Jokers have been removed. Be prepared to explain your reasoning and justify your answers to the class.

1. P(heart) =

2. P(even-numbered card) =

3. P(7 of clubs) =

4. P(1) =

5. P(numbered card) =

6. Arrange the probabilities you found in order from least to greatest.

- ✂ - - -

INVESTIGATION 2: DECK OF CARDS

Use your playing cards to determine the probability of each event. Note that Jokers have been removed. Be prepared to explain your reasoning and justify your answers to the class.

1. P(spade) =

2. P(odd-numbered card) =

3. P(queen of hearts) =

4. P(heart, club, spade, or diamond) =

5. P(Joker) =

6. Arrange the probabilities you found in order from least to greatest.

- ✂ - - -

-- ✂ ----

INVESTIGATION 3: DECK OF CARDS

Use your playing cards to determine the probability of each event. Note that Jokers have been removed. Be prepared to explain your reasoning and justify your answers to the class.

1. $P(11) =$

2. $P(\text{club}) =$

3. $P(\text{face card}) =$

4. $P(\text{ace of diamonds}) =$

5. $P(\text{spade or heart}) =$

6. Arrange the probabilities you found in order from least to greatest.

-- ✂ ----

INVESTIGATION 4: NUMBER CUBE

Use your number cube to determine the probability of each event. Be prepared to explain your reasoning and justify your answers to the class.

1. $P(2) =$

2. $P(\text{odd number}) =$

3. $P(\text{multiple of 3}) =$

4. $P(7) =$

5. $P(\text{prime or composite number}) =$

6. Arrange the probabilities you found in order from least to greatest.

-- ✂ ----

INVESTIGATION 5: NUMBER CUBE

Use your number cube to determine the probability of each event. Be prepared to explain your reasoning and justify your answers to the class.

1. $P(1) =$

2. $P(\text{even number}) =$

3. $P(1, 2, 3, 4, 5, \text{or } 6) =$

4. $P(\text{number greater than 2}) =$

5. $P(\text{two-digit number}) =$

6. Arrange the probabilities you found in order from least to greatest.

-- ✂ ----

Statistics and Probability: 7.SP.6

"Investigate chance processes and develop, use, and evaluate probability models."

6. "Approximate the probability of a chance event by collecting data on the chance process that produces it and observing its long-run relative frequency, and predict the approximate relative frequency given the probability."

BACKGROUND

Theoretical probability can be used to determine the probability of an event after hundreds of trials. To find the theoretical probability of an event, write the ratio of the number of favorable outcomes to the total number of outcomes. Use this ratio to determine the probability after a certain number of trials by writing a proportion.

For example, the theoretical probability of rolling a 1 on a number cube labeled with the numbers 1, 2, 3, 4, 5, and 6 is $\frac{1}{6}$.

To determine the number of times a 1 would be rolled after 300 trials, write the proportion: $\frac{1}{6} = \frac{x}{30}$. Solving this proportion indicates that a 1 would be rolled 50 times out of 300 trials. However, it may not in fact be rolled exactly 50 times. Because this is theoretical probability, the actual outcome may not occur the number of times indicated by theory.

To determine the exact number of times a 1 would be rolled, the number cube would have to be rolled 300 times and the results recorded. This is called experimental probability, the ratio of the number of event occurrences to the total number of trials. As the number of trials increases, the experimental probability approaches the theoretical probability.

 ACTIVITY: PROBABILITY SIMULATIONS

Students will simulate probability situations at a Web site. They will investigate the probability of an event as the number of trials gets larger and use that information to predict probabilities.

MATERIALS

Computers with Internet access.

1. Instruct your students to go to the Web site http://www.mathsonline.co.uk /nonmembers/resource/prob/index.html. Direct your students to the first two activities, "Tossing Coins" and "Spinning Spinners," which work best for this activity, with "Tossing Coins" being the more basic of the two.

2. Instruct your students to experiment with "Tossing Coins" first. Their goal is to investigate what happens to the probability as more trials are added. Note that students may toss 1 to 10 coins, starting with at least 10 trials. Suggest that they begin with 1 coin, for 10 trials, then 20 trials, 30 trials, 50 trials, 100 trials, or more. They should record the number of heads thrown, number of tails thrown, and the number of trials. By keeping a record, students will be able to identify patterns that emerge. If time permits, they can experiment with tossing more than one coin.

3. "Spinning Spinners" is a more complex activity as two spinners are spun at the same time with their results being added or subtracted. You may suggest that students work on this after they have completed investigating "Tossing Coins." Students should try various spinner combinations, record their results, and look for emerging patterns as the number of trials increases.

CLOSURE

Discuss your students' results as a class. Ask questions such as the following: What, if any, patterns did they find as they increased the number of trials? Why were their trials an example of experimental probability? What happened as the number of trials grew larger?

Statistics and Probability: 7.SP.7

"Investigate chance processes and develop, use, and evaluate probability models."

7. "Develop a probability model and use it to find probabilities of events. Compare probabilities from a model to observed frequencies; if the agreement is not good, explain possible sources of the discrepancy.

   a. "Develop a uniform probability model by assigning equal probability to all outcomes, and use the model to determine probabilities of events.

   b. "Develop a probability model (which may not be uniform) by observing frequencies in data generated from a chance process."

BACKGROUND

When finding probability, most students assume that all possibilities are equally likely. However, in many situations, the outcomes are not all equally likely. For example, if students want to find the probability of a soda can landing on its side when tossed, many will assume that the probability is $\frac{1}{3}$ because of three possible outcomes: landing on its top, bottom, or side. However, these three outcomes are not equally likely. Because the side of the can has a larger surface area than the top or bottom, the soda can is more likely to land on its side. In this situation, the theoretical probability cannot be found. The only way to determine its approximate probability is by conducting an experiment and analyzing the results.

 ACTIVITY 1: SPINNER EXPERIMENT

Working in pairs or groups of three, students will investigate the concepts of theoretical versus experimental probability. On a Web site they will spin an interactive spinner, with the on-site computer recording their results. They may adjust the number of sections on the spinner to examine the effects of more or fewer outcomes.

MATERIALS

Computers with Internet access.

PROCEDURE

1. Instruct your students to go to the Web site http://www.shodor.org/interactivate/activities/BasicSpinner/. The spinner on the Web site begins with four equal sectors, each with a different color. Ask your students to predict the probability of each color sector occurring with each spin. It is $\frac{1}{4}$.

2. Instruct your students to spin the spinner. Note that their results will be automatically recorded. After several spins students should write their results on a sheet of paper. By clicking on the "Show results frame," a graph of the experimental outcomes will be displayed.

3. Instruct your students to add another sector to the spinner and repeat the same process. After several spins, again have them record their results on a sheet of paper. Instruct students to click on the "Show results frame."

4. Finally, have your students select another number of sectors for their spinner and repeat the experiment. Have them record their results on a sheet of paper. Instruct students to click on the "Show results frame."

CLOSURE

Discuss the results of each experiment, beginning with the spinner that had four sectors. Discuss what the students estimated each probability to be. For the spinner with four sectors, the theoretical probability is $\frac{1}{4}$ for each color. For the spinner with five sectors, the theoretical probability is $\frac{1}{5}$ for each sector, and so on because each sector on the spinner is the same size. Therefore, all outcomes are equally likely. Ask your students to share the results of their experiments. It is likely that the results of the experiment may not match the theoretical probabilities, especially if the students did not complete many trials.

ACTIVITY 2: FLIPPING A MARSHMALLOW

Students will work in pairs or groups of three to flip a marshmallow and record the results. They will use their information to determine experimental probability.

MATERIALS

Two bags of large marshmallows, at least one marshmallow for each pair or group of students; thin dark marker.

Label each marshmallow with a "T" for top and "B" for bottom.

PROCEDURE

1. Hold up a marshmallow. Tell your students that they are going to conduct an experiment in which they flip a marshmallow to determine how it lands: on its top, bottom, or side. Before beginning the experiment, have your students predict the probability that the marshmallow will land on its top. (Caution students that they are not to eat the marshmallows.)

2. Distribute the marshmallows. Explain that students are to flip the marshmallow 50 times and record, on a sheet of paper, how it lands each time. (The number of trials can be reduced to 25 or expanded to 80 or 100, depending on how much time you have in class.)

3. Once students have completed the experiment, write the class data on the board.

4. Ask your students to use the class data to make conjectures about this experiment. What are some patterns they noticed? Students should notice the marshmallow landed on its side more than on its top or bottom. Discuss that this occurred because the side of the marshmallow has a larger surface area and the marshmallow therefore is more likely to land on its side. Have students find the experimental probability of the data in their pairs or groups, as well as in the class data. Note that experimental probability is found by writing the ratio of the number of desired outcomes to the total number of trials. Discuss if the results of the partners or groups were similar to the class data. Discuss any discrepancies that occurred. Because this is an experiment, the results are based upon what happened in each group. Not every group will have the exact same data.

CLOSURE

Have your students discuss, in pairs or groups, how the class data compared to their original prediction of the probability. Why might there be a discrepancy? Discuss the answers as a class. (Answers may vary, but students should realize that their original predictions were examples of theoretical probability, which may not always match the results of experimental probability, particularly if there are a limited number of outcomes.)

Statistics and Probability: 7.SP.8

"Investigate chance processes and develop, use, and evaluate probability models."

8. "Find probabilities of compound events using organized lists, tables, tree diagrams, and simulation.

   a. "Understand that, just as with simple events, the probability of a compound event is the fraction of outcomes in the sample space for which the compound event occurs.

   b. "Represent sample spaces for compound events using methods such as organized lists, tables, and tree diagrams. For an event described in everyday language, identify the outcomes in the sample space which compose the event.

   c. "Design and use a simulation to generate frequencies for compound events."

BACKGROUND

A compound event is the outcome of a probability experiment that involves more than one event. It contrasts with simple probability, which is the probability of one event occurring. For example, rolling one number cube is an example of simple probability. Rolling one die and then another is an example of compound probability. Compound probability can be found by writing the ratio of the number of favorable outcomes to the total number of outcomes. A list of all of the possible outcomes is called a sample space. A sample space can be represented by organized lists, tables, and tree diagrams.

 ### ACTIVITY 1: WORKING WITH SAMPLE SPACES

Working in groups of four or five, students will use an organized list, table, or tree diagram to create a sample space. Using their representations, students will solve compound probability problems.

MATERIALS

Transparencies, one per group; markers; tissues or erasers.

PROCEDURE

1. Each group will use one of three different methods for completing this activity. If your class is relatively small, divide students into three groups. Divide larger classes into six groups.

2. Assign each group one of the following ways to organize possible outcomes: an organized list, table, or tree diagram.

3. Ask your students to imagine that one die is rolled and then another is rolled. Instruct the groups to represent the sample space, using the method assigned to them. The sample space should contain 36 outcomes: (1, 1), (1, 2), (1, 3) . . . (6, 4), (6, 5), (6, 6).

4. Instruct each group to write their sample space on a transparency.

5. Ask students to use their sample space to list the outcomes described below and write the answers on their transparency.

   - Rolling a double. Student results should be (1, 1), (2, 2), (3, 3), (4, 4), (5, 5), and (6, 6).

   - Rolling a sum of 10. Student results should be (4, 6), (5, 5), and (6, 4).

   - Rolling two even numbers. Student results should be (2, 2), (2, 4), (2, 6), (4, 2), (4, 4), (4, 6), (6, 2), (6, 4), and (6, 6).

6. Instruct students to use their sample spaces to find the probability of the events listed above by writing the ratio of the number of favorable outcomes to the total number of outcomes. (The answers are $\frac{6}{36} = \frac{1}{6}$; $\frac{3}{36} = \frac{1}{12}$; $\frac{9}{36} = \frac{1}{4}$.) Students should then write their answers on their transparency.

CLOSURE

Have the groups display their sample spaces using an overhead projector. Each group should explain the method they used to represent the sample space. Discuss the probability that each group determined. Discuss which method students found easiest to use. Did different methods work better for different problems? Why or why not?

 ## ACTIVITY 2: SIMULATING EVENTS

Students will work in groups of four or five to design a simulation involving compound events. They will use this simulation to find the frequencies and probabilities of the compound events they designed.

PROCEDURE

1. Explain that students will design a simulation involving compound events. Note that they used a simulation in the previous activity to represent the sample space for rolling one die and then another.

2. Note that this time students will design a simulation of their own by using compound events. Provide some examples such as:

- Spinning a spinner, then tossing a die.

- Selecting a letter from a box and tossing a coin.

- Pulling names from a hat and spinning a spinner.

3. Explain that students must write a description of the events. For example, if students choose to use a spinner for their simulation, they must describe the numbers on the spinner and the number of equally divided parts of the spinner. If they choose to select a letter from a box, they must state which letters and how many of each will be in the box.

4. After students describe the events, they should represent the sample space, using an organized list, table, or tree diagram. Encourage them to double-check their work for accuracy and correct any errors.

5. Explain that students are to now write questions based on the events they described, and that these questions will be answered by another group. Following is an example of a possible question: Find the probability of spinning a 4 and selecting an A from a hat. Note that students are to write their answers on a separate sheet of paper.

6. Have students exchange papers with another group; groups will then answer each other's questions. They should conference to resolve any disputed results.

CLOSURE

Discuss the simulations. Which simulations were most interesting? Were any confusing? What are the components of a good simulation?

Standards and Activities for Grade 8

The Number System: 8.NS.1

"Know that there are numbers that are not rational, and approximate them by rational numbers."

> 1. "Know that numbers that are not rational are called irrational. Understand informally that every number has a decimal expansion; for rational numbers show that the decimal expansion repeats eventually, and convert a decimal expansion which repeats eventually into a rational number."

BACKGROUND

Rational numbers can be expressed as $\frac{a}{b}$, where a and b are integers, $b \neq 0$. All rational numbers have a terminating or repeating decimal expansion, which can be found by dividing the numerator by the denominator. (Irrational numbers have a decimal expansion that neither terminates nor repeats.)

To convert a decimal expansion that terminates to a rational number, write the decimal as a rational number. Simplify, if possible.

To convert a decimal expansion that repeats to a rational number, do the following:

- Let $n =$ the repeating decimal.

- If one digit repeats, multiply the original equation by 10. If two digits repeat, multiply the original equation by 100. If three digits repeat, multiply the original equation by 1,000.

- Subtract the original equation from the transformed equation. Be sure to align the decimal points.

- Solve for x.

- Simplify if possible.

Examples:

$$
\begin{aligned}
n &= 0.\overline{4} \\
10n &= 4.\overline{4} \\
-n &= 0.\overline{4} \\
\hline
9n &= 4 \\
n &= \frac{4}{9}
\end{aligned}
\qquad\qquad
\begin{aligned}
n &= 0.\overline{45} \\
100n &= 45.\overline{45} \\
-n &= 0.\overline{45} \\
\hline
99n &= 45 \\
n &= \frac{45}{99} = \frac{5}{11}
\end{aligned}
$$

ACTIVITY: MATH TIC-TAC-TOE

In pairs or groups of three, students will play a game of tic-tac-toe. They will be presented with ten decimals and two tic-tac-toe boards that they will copy. They will express the decimals as rational numbers.

PROCEDURE

1. Present the ten decimals and the two tic-tac-toe boards shown below to your students. Instruct students to copy the numbers and the tic-tac-toe boards.

1) $0.\overline{3}$ 2) $0.\overline{27}$ 3) $0.\overline{1}$ 4) $0.\overline{16}$ 5) 0.1 6) $0.\overline{6}$ 7) $0.8\overline{3}$ 8) 0.5 9) $0.\overline{23}$ 10) $0.\overline{36}$

$\frac{1}{9}$	$\frac{3}{11}$	$\frac{6}{7}$	$\frac{1}{13}$	$\frac{1}{10}$	$\frac{1}{7}$
$\frac{8}{11}$	$\frac{23}{99}$	$\frac{1}{2}$	$\frac{25}{99}$	$\frac{2}{11}$	$\frac{5}{6}$
$\frac{3}{22}$	$\frac{2}{3}$	$\frac{2}{7}$	$\frac{1}{6}$	$\frac{4}{11}$	$\frac{1}{3}$

2. Instruct your students to convert each decimal to a rational number. If the rational number is on the tic-tac-toe board, students should "X" it out. The first person in each group to get three X's in a row, column, or along a diagonal of each board is the winner.

3. If time permits, encourage your students to create a math tic-tac-toe game of their own.

CLOSURE

Provide the correct solutions to the problems.

ANSWERS

The second column of the first board should have three X's, as should the last row of the second board. The rational numbers that correspond to each repeating decimal follow:

1) $\frac{1}{3}$ 2) $\frac{3}{11}$ 3) $\frac{1}{9}$ 4) $\frac{1}{6}$ 5) $\frac{1}{10}$ 6) $\frac{2}{3}$ 7) $\frac{5}{6}$ 8) $\frac{1}{2}$ 9) $\frac{23}{99}$ 10) $\frac{4}{11}$

The Number System: 8.NS.2

"Know that there are numbers that are not rational, and approximate them by rational numbers."

> 2. "Use rational approximations of irrational numbers to compare the size of irrational numbers, locate them approximately on a number line diagram, and estimate the value of expressions."

BACKGROUND

Irrational numbers are numbers that cannot be expressed as $\frac{a}{b}$ where a and b are integers, $b \neq 0$. Unlike rational numbers whose decimal equivalents either terminate or repeat in a specific pattern, the decimal equivalents of irrational numbers can only be approximated. For example, the irrational number $\sqrt{2}$ is about 1.41, rounded to the nearest hundredth. This can be written as $\sqrt{2} \approx 1.41$. ($\approx$ means "approximately equal to"). Because finding a square root of a number is the opposite of squaring a number, the estimated value of a square root can be checked by squaring a number.

 ACTIVITY 1: ZEROING IN

Working in pairs or groups of three, students will approximate the values of irrational numbers and graph them on the number line.

MATERIALS

Graph paper ($\frac{1}{4}''$ grid); rulers.

PROCEDURE

1. Demonstrate how students may use guess and check to find an approximate value of $\sqrt{2}$.

   - Ask your students what number squared is approximately equal to 2. For example, 1.5 is a possible response.

   - Ask your students to check their answer by squaring it. Some may find that their answer is more than 2; others may find that it is less than 2.

   - Use 1.5 as an example and square it for your students. $1.5^2 = 2.25$, which is greater than 2.

   - Suggest a smaller estimate, perhaps 1.3. Square 1.3. $1.3^2 = 1.69$, which is less than 2.

- Continue this process until $\sqrt{2}$ is found to be a number between two consecutive digits in the hundredths place. One estimate should be greater than the radicand, and the other should be less than the radicand. It may be helpful to organize the approximations in tables, such as those shown below.

Estimate	Square		Too Small	Too Large
1.5	$1.5^2 = 2.25$		1.3	1.5
1.3	$1.3^2 = 1.69$		1.4	1.43
1.4	$1.4^2 = 1.96$		1.41	1.42
1.43	$1.43^2 = 2.0449$			
1.42	$1.42^2 = 2.0164$			
1.41	$1.41^2 = 1.9881$			

$\sqrt{2}$ is between 1.41 and 1.42.

2. Instruct your students to work in pairs or groups to find $\sqrt{3}$, $\sqrt{5}$, $\sqrt{7}$, and $\sqrt{8}$. They should record their estimates, square them, and organize their data in a table.

3. Correct your students' estimates. $\sqrt{3}$ is between 1.73 and 1.74; $\sqrt{5}$ is between 2.23 and 2.24; $\sqrt{7}$ is between 2.64 and 2.65; and $\sqrt{8}$ is between 2.82 and 2.83. Note that all of the estimates are between 1 and 3.

4. Instruct your students to graph the approximate values on the number line. Depending upon the abilities of your students, you may let them select a scale; or suggest that they hold their graph paper horizontally and draw a number line so that every two units on the graph is equal to 0.1.

CLOSURE

Check your students' graphs. Ask them why they did not have to find approximations for $\sqrt{1}$, $\sqrt{4}$, and $\sqrt{9}$. The answer, of course, is that each is a rational number, and they have exact values, 1, 2, and 3, respectively.

ACTIVITY 2: IRRATIONAL NUMBERS—THEY'RE INSANE

Students will write a story or a paragraph explaining how they think irrational numbers got to be called "irrational numbers."

PROCEDURE

1. Share the dictionary definition of "irrational" with your students. Irrational—incapable of exercising the power of reason; absurd.

2. Provide some background about the ancient Greeks who were first-class mathematicians (think Pythagoras), but who believed that all numbers were counting numbers or fractions. Ask your students to imagine the Greeks' surprise when a solution to an equation turned out to be $x = \sqrt{2}$.

3. Using this information and a bit of imagination and creativity, ask your students to write a story or paragraph about how numbers such as $\sqrt{2}$ acquired the name of irrational numbers.

CLOSURE

Share the stories of your students. You might have students read excerpts of them or you might display the stories in the classroom.

Expressions and Equations: 8.EE.1

"Work with radicals and integer exponents."

1. "Know and apply the properties of integer exponents to generate equivalent numerical expressions."

BACKGROUND

An exponent states the number of times a base is used as a factor. For example, 2^3 can be rewritten as $2 \cdot 2 \cdot 2$ or 8.

The properties of exponents help students simplify and generate equivalent expressions. The properties of exponents are listed below:

- Product of Powers Property: $x^m x^n = x^{m+n}$. To multiply two powers having the same base, add the exponents.

- Power of a Power Property: $(x^m)^n = x^{mn}$. To find the power of a power, multiply the exponents.

- Power of a Product Property: $(xy)^m = x^m y^m$. To find the power of a product, find the power of x and y and multiply.

- Quotient of Powers Property: $\dfrac{x^m}{x^n} = x^{m-n}, x \neq 0$. To divide two powers having the same base, subtract the exponents.

- Power of a Quotient Property: $\dfrac{x^m}{y^m} = \left(\dfrac{x}{y}\right)^m, y \neq 0$. To find the power of a quotient, find the power of x and y and divide.

- Zero Exponent Property: $x^o = 1, x \neq 0$. Any number, except 0, that is raised to the 0 power is equal to 1.

- Negative Exponent Property: $x^{-m} = \dfrac{1}{x^m}$ and $\dfrac{1}{x^{-m}} = x^m, x \neq 0$. A number raised to a negative power is equal to 1 over the same base raised to the positive power.

 ACTIVITY: WHAT DOES IT EQUAL?

Students will be given a slip of paper that contains a question and a statement regarding an expression. Working individually or in pairs, students will identify equivalent expressions.

MATERIALS

Reproducible, "Equivalent Expressions."

You may prefer to enlarge the reproducible before photocopying. Cut out each box so that you have a total of 21 slips of paper. Keep a copy for yourself to refer to during the activity. Note that the slips are arranged in order on the reproducible, each providing the correct answer to the question written on the preceding slip. (The first answer, "It equals $\frac{1}{6}$," is the answer to the last question on the reproducible.)

PROCEDURE

1. Before passing out the slips of paper to your students, mix the slips up.

2. Distribute one slip of paper to each student (or a slip to pairs of students). For a small class, you may give some students two slips. You must distribute all 21 slips.

3. To start, choose a student to read the question written on the right side of his slip. You may find it helpful to write the expression on the board or project it on a screen. All students should try to find an equivalent expression on the left side of their slips. Because of the way the slips are designed, only one will contain the correct equivalent expression. The student who has the slip with the correct equivalent expression should say "It equals. . ." and then provide the answer. If the student is correct, he then reads the question written on the right side of his slip. If he is incorrect, point out his error. Another student should then provide the correct equivalent expression from the left side of her slip.

4. Continue the process until the student who read the first question has the correct response to the last question.

CLOSURE

Discuss the activity. Did students find other answers in addition to those on the cards? Reinforce the idea that one expression may be written in several ways.

EQUIVALENT EXPRESSIONS

It equals $\dfrac{1}{6}$	What does $3^2 \cdot 3^5$ equal?		It equals 3^7	What does 2^3 equal?		It equals 8	What does $\dfrac{3^5}{3^2}$ equal?

It equals 3^3	What does 2^2 equal?		It equals 4	What does $\dfrac{1}{10^2}$ equal?		It equals 10^{-2}	What does 6^2 equal?

It equals 36	What does 5^{-2} equal?		It equals $\dfrac{1}{25}$	What does 10^0 equal?		It equals 1	What does 2^1 equal?

It equals 2	What does $(2 \cdot 4)^2$ equal?		It equals $2^2 \cdot 4^2$	What does $\left(\dfrac{2}{5}\right)^3$ equal?		It equals $\dfrac{8}{125}$	What does $\dfrac{1}{3^3}$ equal?

It equals $\dfrac{1}{27}$	What does $(2^2)^3$ equal?		It equals 2^6	What does 8^{-1} equal?		It equals $\dfrac{1}{8}$	What does $2^5 \cdot 2^{-10}$ equal?

It equals 2^{-5}	What does $10^2 \cdot 10^{-1}$ equal?		It equals 10	What does $(2 \cdot 3)^3$ equal?		It equals $2^3 \cdot 3^3$	What does $\left(\dfrac{3}{7}\right)^2$ equal?

It equals $\dfrac{3^2}{7^2}$	What does $\dfrac{5^2}{5^{-3}}$ equal?		It equals 5^5	What does $\dfrac{1}{3^{-1}}$ equal?		It equals 3	What does $(2 \cdot 3)^{-1}$ equal?

Expressions and Equations: 8.EE.2

"Work with radicals and integer exponents."

2. "Use square root and cube root symbols to represent solutions to equations of the form $x^2 = p$ and $x^3 = p$, where p is a positive rational number. Evaluate square roots of small perfect squares and cube roots of small perfect cubes. Know that $\sqrt{2}$ is irrational."

BACKGROUND

Finding the square root of a number is the opposite (or inverse) of squaring a number. For example, $\sqrt{36} = 6$, which is read "the square root of 36 equals 6," because $6^2 = 36$. In general, $\sqrt{a} = b$, if $b^2 = a$.

The square roots of perfect squares are rational numbers. The square roots of numbers that are not perfect squares are irrational numbers. A list of the numbers from 1 through 12, their squares, and the square roots of the squares is shown below.

Number	1	2	3	4	5	6	7	8	9	10	11	12
Square	1	4	9	16	25	36	49	64	81	100	121	144
Square Root	1	2	3	4	5	6	7	8	9	10	11	12

To represent solutions to equations of the form $x^2 = p$, rewrite the equation as $x = \sqrt{p}$. For example, to represent the solutions of $x^2 = 36$, rewrite the equation as $x = \sqrt{36}$.

Finding the cube root of a number is the opposite (or inverse) of cubing a number. For example, $\sqrt[3]{1,000} = 10$, which is read "the cube root of 1,000 equals 10," because $10^3 = 1,000$. In general, $\sqrt[3]{a} = b$, if $b^3 = a$.

The cube roots of perfect cubes are rational numbers. The cube roots of numbers that are not perfect cubes are irrational numbers. A list of the numbers from 1 through 12, their cubes, and the cube roots of the cubes is shown below.

Number	1	2	3	4	5	6	7	8	9	10	11	12
Cube	1	8	27	64	125	216	343	512	729	1000	1331	1728
Cube Root	1	2	3	4	5	6	7	8	9	10	11	12

To represent solutions to equations of the form $x^3 = p$, rewrite the equation as $x = \sqrt[3]{p}$. For example, to represent the solutions of the equation $x^3 = 1,000$, rewrite the equation as $x = \sqrt[3]{1,000}$.

ACTIVITY: CREATING SQUARES AND CUBES

Working in groups of three or four, students will create squares and cubes to geometrically model perfect squares and perfect cubes. They will use the formula for finding the area of a square, $A = s^2$, and rewrite the formula by expressing it as $s = \sqrt{A}$. They will use the formula for finding the volume of a cube, $V = e^3$, and rewrite the formula by expressing it as $e = \sqrt[3]{V}$.

MATERIALS

About 50 commercially produced cubes, such as MathLink® Cubes, Multilink Cubes, or Centimeter Cubes, for each group.

PROCEDURE

1. Distribute the cubes to each group of students.

2. Instruct your students to use the faces (or tops) of the cubes to form as many larger squares as they can. They should consider only the length and width of the cubes, not the height. Students should record the number of faces that form a square and the length of each side. For example, four faces form a square that has an area of 4 square units; the length of each side is $\sqrt{4}$ or 2 units. Note that the area of each square is a square of a whole number, which is called a perfect square. The length of the side of the square is the square root of the area. Ask your students to write the formulas for finding the area of each of the squares they formed in terms of s, where s is the length of the square.

3. Next instruct your students to use their cubes to form as many larger cubes as they can. Students should record the number of cubes that form each larger cube and the length of each side of the face, which is called the edge. For example, eight cubes form a larger cube that has a volume of 8 cubic units; each edge is $\sqrt[3]{8}$ or 2 units long. Note that the volume of a cube is a cube of a whole number, which is called a perfect cube. The length of the edge is the cube root of the volume. Ask your students to write formulas for finding the volume of the cubes they formed in terms of e, where e is the length of the edge of the cube.

CLOSURE

Check your students' answers. $4 = s^2, s = \sqrt{4}$; $9 = s^2, s = \sqrt{9}$; $16 = s^2, s = \sqrt{16}$; $25 = s^2, s = \sqrt{25}$; $36 = s^2, s = \sqrt{36}$; $49 = s^2, s = \sqrt{49}$; $8 = e^3, e = \sqrt[3]{8}$; $27 = e^3, e = \sqrt[3]{27}$. Ask your students to offer examples of other perfect squares and their square roots, and examples of other perfect cubes and their cube roots.

Expressions and Equations: 8.EE.3

"Work with radicals and integer exponents."

3. "Use numbers expressed in the form of a single digit times an integer power of 10 to estimate very large or very small quantities, and to express how many times as much one is than the other."

BACKGROUND

Numbers expressed in the form of a number greater than or equal to 1 and less than 10 times an integer power of 10 are said to be written in scientific notation. For example, $2.4 \times 10^2 = 240$, $1.256 \times 10^{-2} = 0.01256$, and $3.79 \times 10^5 = 379,000$.

Students can estimate very large or very small quantities that are expressed in scientific notation. Thus $240 = 2.4 \times 10^2$ or about 2×10^2; $0.01256 = 1.256 \times 10^{-2}$ or about 1×10^{-2}; and $379,000 = 3.79 \times 10^5$ or about 4×10^5. Once students have made their estimates, they can express how many times one number is than another by using the quotient of powers property of exponents: $\dfrac{x^m}{x^n} = x^{m-n}, x \neq 0$.

 ACTIVITY 1: EXPRESSING NUMBERS IN SCIENTIFIC NOTATION

Working at a Web site, students will practice converting very large and very small numbers into scientific notation.

MATERIALS

Computers with Internet access.

PROCEDURE

1. Instruct your students to go to the Web site www.xpmath.com/ and click on "Algebra." Under Game Search, search for scientific notation, then click on "converting large numbers into scientific notation," or "converting small numbers into scientific notation."

2. Explain that during a 60-second time period, students will be presented with numbers that they must convert into scientific notation. By clicking on "Check," they will immediately find that their answer is either correct or incorrect. They will then be presented with another problem to convert. This process continues until the time period is up. Students will then have the opportunity to correct their mistakes.

3. Note that students may repeat the activity for additional practice. Encourage students who previously converted large numbers into scientific notation to now convert very small numbers, and encourage those who previously converted small numbers to convert large numbers.

CLOSURE

Discuss the activity. Ask your students to explain the usefulness of using scientific notation. Students should support their answers.

 ## ACTIVITY 2: IT'S HOW MANY TIMES AS MUCH...

Working in pairs or groups of three, students will select a topic, gather data, express the data in scientific notation, estimate the numbers, and compare data.

MATERIALS

Computers with Internet access; almanacs or similar references. Optional: markers; rulers; poster paper.

PROCEDURE

1. Explain that your students are to select a topic, such as the highest-grossing films, the number of cell phone users or Internet users, population (of a country or state), the solar system, the number of cells in organisms, the relative sizes of atoms, computer memory, or any other topic where the numbers are very large or very small.

2. Instruct them to find at least five numbers related to their topic. They should cite their source. Students are to first express these numbers in scientific notation, and then estimate the number.

3. Explain that once they have estimated their numbers, they are to determine about how many times as much one quantity is than another.

4. You might want to provide your students with an example of what you expect them to do. According to the *Time for Kids Almanac 2011*, the diameter of Mars is 4,222 miles or about 4×10^3 miles. The diameter of Jupiter is 88,650 miles or about 9×10^4 miles. Students may use their estimates to conclude that the diameter of Jupiter is a little more than 20 times the diameter of Mars. One way to make this estimate follows:

$$\frac{9 \times 10^4}{4 \times 10^3} = 2.25 \times 10 = 22.5 \text{ or about } 20$$

5. Instruct your students to display their findings in the form of a chart or table. If your classroom has the necessary computer equipment, consider having your students create their charts using computers and then project their charts on the screen. An option is for students to make charts using conventional materials.

CLOSURE

Have students share their charts with the class or display them around the room. Ask your students the following questions: How does rounding affect estimates? Why are some numbers easier to work with than others? Do you feel that estimating with scientific notation is a useful strategy? What are its advantages and disadvantages?

Expressions and Equations: 8.EE.4

"Work with radicals and integer exponents."

4. "Perform operations with numbers expressed in scientific notation, including problems where both decimal and scientific notation are used. Use scientific notation and choose units of appropriate size for measurements of very large or very small quantities. Interpret scientific notation that has been generated by technology."

BACKGROUND

To multiply numbers expressed in scientific notation, do the following:

- Multiply each of the numbers that are multiplied by a power of 10.

- Use the product of powers property, $x^m \cdot x^n = x^{m+n}$, which states that if the bases are the same, the exponents must be added.

- Express the product in scientific notation.

  Example: $(3.4 \times 10^3)(8.2 \times 10^5) = 27.88 \times 10^8 = 2.788 \times 10^9$

To divide numbers expressed in scientific notation, do the following:

- Divide each of the numbers that are multiplied by a power of 10.

- Use the quotient of powers property, $\dfrac{x^m}{x^n} = x^{m-n}, x \neq 0$, which states that if the bases are the same, the exponents must be subtracted.

- Express the quotient in scientific notation.

  Example: $\dfrac{(2.6 \times 10^8)}{5 \times 10^3} = 0.52 \times 10^5 = 5.2 \times 10^4$

To add or subtract numbers expressed in scientific notation, do the following:

- Rewrite each number so that the bases have the same power of 10.

- Add or subtract each of the numbers that are multiplied by 10.

- Express the sum or difference in scientific notation.

  Example: $3.8 \times 10^3 + 2.85 \times 10^2 = 38 \times 10^2 + 2.85 \times 10^2 = 40.85 \times 10^2 = 4.085 \times 10^3$

If a decimal is not expressed in scientific notation, students should rewrite it in scientific notation. Note that there are other notations that have been generated by technology. For example, 3×10^8 may be written as 3E8 or $3 * 10 \wedge 8$.

ACTIVITY 1: MATH BINGO

Students will play Math Bingo, a game similar to traditional bingo. They will be given a board and problems involving a product, quotient, sum, or difference of two numbers expressed in scientific notation. They will locate their answers on the board.

MATERIALS

Reproducible, "Math Bingo Board."

PROCEDURE

1. Distribute a copy of the Math Bingo Board to each student.

2. Explain that they should randomly write one number or power of 10 found below their board in the boxes, one per box. (These numbers will be the answers to the problems you will present; however, not all of the numbers will be used as answers.)

3. After students have filled in the boxes on their boards, present the problems below:

   1. $2,850 \times (3.6 \times 10^2)$

   2. $-0.013 \times (2.8 \times 10^{-1})$

   3. $(2.19 \times 10^{-3}) \times (1.7\ E2)$

   4. $(-1.08 \times 10^2) \div (6 \times 10^{-2})$

   5. $(4.76 \times 10^3) \div (2.5 \times 10^{\wedge}-1)$

   6. $(1.32 \times 10^{-2}) + (5 \times 10^{-2})$

   7. $(3 \times 10^{-5}) - (2 \times 10^{\wedge}-4)$

   8. $1.6 - (2 \times 10^{-3})$

4. Instruct your students to solve the problems, expressing their answers in scientific notation.

5. Students should then locate their answers on the Math Bingo Board and cross them out. Emphasize that each of the correct answers should have a number that appears on one space and a "times a power of 10" on another space. For example, if an answer were 2.3×10^5, students would find 2.3 on one space and cross it out, and $\times 10^5$ on another space and cross it out. The first student to cross out a row, column, or diagonal wins.

CLOSURE

Provide the correct solution to each problem. Answer any questions your students may have.

ANSWERS

1) 1.026×10^6 **2)** -3.64×10^{-3} **3)** 3.723×10^{-1} **4)** -1.8×10^3

5) 1.904×10^4 **6)** 6.32×10^{-2} **7)** -1.7×10^{-4} **8)** 1.598×10^0

ACTIVITY 2: METRIC PREFIXES

Working in small groups, students will compile a list of metric prefixes, their meanings, and examples of things measured, using these prefixes.

MATERIALS

Math texts; math reference books. Optional: computers with Internet access.

PROCEDURE

1. Divide your students into eight groups.

2. Assign each group one of the following metric prefixes: peta, tera, gigag, mega, micro, nana, pico, femto.

3. Explain that each group will research their prefix, state its meaning, and list examples of things that are measured with it. Each group will decide upon a spokesperson, who will present the group's findings to the class.

CLOSURE

Set aside some class time for students to share their results.

		Free Space		

Use the numbers below to fill in the spaces on your board.

-3.64	-2.76	-1.8	-1.7	1.026	1.598	1.8	1.904
3.723	6.32	6.34	$\times 10^{-5}$	$\times 10^{-4}$	$\times 10^{-3}$	$\times 10^{-2}$	$\times 10^{-1}$
$\times 10^{0}$	$\times 10^{1}$	$\times 10^{2}$	$\times 10^{3}$	$\times 10^{4}$	$\times 10^{5}$	$\times 10^{6}$	$\times 10^{7}$

Expressions and Equations: 8.EE.5

"Understand the connections between proportional relationships, lines, and linear equations."

> 5. "Graph proportional relationships, interpreting the unit rate as the slope of the graph. Compare two different proportional relationships represented in two different ways."

BACKGROUND

Two quantities vary directly if there exists a number k such that $y = kx, k \neq 0$. k is called the constant of variation. The graph of $y = kx$ is a line through the origin of the coordinate plane.

Because slope is defined as the number of units a line rises (or falls) for each unit of horizontal change, and a unit rate is the rate per one given unit, the unit rate is the same as the slope of a non-vertical line.

One example of two quantities that vary directly is the diameter and radius of a circle. $d = 2r$. The graph of $d = 2r$ is a line through the origin. The unit rate and the slope of the line are both 2.

 ## ACTIVITY: IT'S PROPORTIONAL

Students will graph proportional relationships by hand and by using a virtual graphing calculator.

MATERIALS

Graph paper; rulers; computers with Internet access; digital projector.

PROCEDURE

1. Instruct your students to graph the equation $y = 2x$ by finding ordered pairs that are solutions to the equation. $(0, 0)$, $(2, 4)$, $(6, 12)$, $(3, 6)$, $(-2, -4)$ are some solutions. Students should then graph each point, noting that all of the points lie on a line. Explain that a line that goes through all of these points is the graph of the equation of the line, $y = 2x$. Note the ratio $\frac{y}{x} = 2$ for each ordered pair.

2. Instruct your students to graph $y = 3x$ and $y = 4x$. Using the same graph paper, they should find and graph the ordered pairs that are solutions to each of these equations. Consider the ratio $\frac{y}{x}$ for each set of points. Ask your students how this ratio compares with the coefficient of x in each equation. (The ratio is 3:1 for the graph of $y = 3x$ and 4:1 for the graph of $y = 4x$.)

3. Go to the Web site http://my.hrw.com/math06_07/nsmedia/tools/Graph_Calculator /graphCalc.html and demonstrate the use of the virtual graphing calculator. Enter an equation such as $y_1 = 2x$ and then click on "Graph" to graph the line. Click "Trace" and then click on the right arrow under "Zoom Out" to display the ordered pairs that lie on the line. Note that in all cases, $\frac{y}{x} = \frac{2}{1}$. For each unit on the x-axis, the y-coordinate doubles, reinforcing the concept of the slope of a line. $m = \dfrac{\text{vertical change}}{\text{horizontal change}}$

4. Leave the graph of $y_1 = 2x$ on the screen and enter $y_2 = 3x$ and $y_3 = 4x$. Follow the same procedure: "Graph," then "Trace" and "Zoom Out."

5. Instruct your students to visit this Web site. They may enter an equation, virtually graph the equation, and make and test conjectures.

CLOSURE

Pose the following to your students: Suppose you are given two different proportional relationships. One is a graph and the other is an equation. How can you tell which has the greater slope? Provide examples to support your answer.

Student answers may vary. Students may represent the graph as an equation, or the equation as a graph, and then compare the slope. The steeper the slope, the greater the rate of change.

Expressions and Equations: 8.EE.6

"Understand the connections between proportional relationships, lines, and linear equations."

6. "Use similar triangles to explain why the slope m is the same between any two distinct points on a non-vertical line in the coordinate plane; derive the equation $y = mx$ for a line through the origin and the equation $y = mx + b$ for a line intersecting the vertical axis at b."

BACKGROUND

The slope, m, of a line is a constant. The slope of a line may be expressed in a variety of ways:
$$m = \frac{\text{rise}}{\text{run}} \text{ or } m = \frac{\text{vertical change}}{\text{horizontal change}} \text{ or } m = \frac{\text{change in } y}{\text{change in } x} \text{ or } m = \frac{y_2 - y_1}{x_2 - x_1}.$$
An equation of a straight line through the origin is $y = mx$. The equation of a line through $(0, b)$ on the y-axis is $y = mx + b$.

ACTIVITY 1: THE SLOPE IS THE SAME

Students will work in pairs or groups of three to show that the slope of a line between any two points on a non-vertical line is the same.

MATERIALS

Graph paper; rulers; reproducible, "Student Guide for the Slope Is the Same."

PROCEDURE

1. Distribute copies of the reproducible, which outlines the steps students should follow in order to complete the activity. Review the steps on the reproducible with your students. Emphasize that the line students draw should be non-vertical and have a positive slope.

2. As students work, offer guidance as necessary. Especially check that students draw similar triangles and that they label the vertices of the triangles correctly.

CLOSURE

Discuss the answers to the questions on the reproducible. Students should realize that the slope of a non-vertical line between any two points on the line is the same, because they showed the slope of the hypotenuse is the same and it is a segment of the line they drew.

Working in pairs or groups of three, students will derive the equations $y = mx$ and $y = mx + b$ by drawing similar triangles, finding the slope of each hypotenuse, and using the equation $m = \dfrac{\text{length of the vertical line segment}}{\text{length of the horizontal line segment}}$.

MATERIALS

Graph paper; rulers; 2-page reproducible, "Student Guide for I Have Derived It."

PROCEDURE

1. Review the properties of similar triangles with your students. Note that similar triangles have the same shape but are not necessarily the same size. Corresponding sides have the same scale factor. Explain that students will use similar triangles to derive the equations $y = mx$ and $y = mx + b$.

2. Review the information on the reproducible with your students. You might ask them to draw similar triangles that have the same slope before using the generalized coordinates. Note that the activity has two parts. Students are to complete both parts.

3. Once students begin working, circulate around the room and monitor their work. Make sure that your students are finding the correct values. In Part 1, the length of the horizontal line segment is x, and the length of the vertical line segment is y. $m = \dfrac{y}{x}$. The equation is $y = mx$. In Part 2, the length of the horizontal line segment is x, and the length of the vertical line segment is $y - b$. $m = \dfrac{y - b}{x}$. The equation is $y = mx + b$.

CLOSURE

Discuss the formulas. Ask your students if they could use the points $(0, 0)$ and (x, y) to derive $y = mx$ and the points $(0, b)$ and (x, y) to derive $y = mx + b$ rather than using the graphs. Encourage your students to support their answers. In both cases, they would be able to derive the same formulas by using the formula $m = \dfrac{y_2 - y_1}{x_2 - x_1}$ instead of drawing similar triangles.

STUDENT GUIDE FOR "THE SLOPE IS THE SAME"

1. Use your ruler to draw the *x*-axis and *y*-axis on your graph paper.

2. Draw a straight, non-vertical line that has a positive slope on your graph.

3. Draw a right triangle so that one leg is parallel to the *y*-axis, one leg is parallel to the *x*-axis, and the hypotenuse lies on the non-vertical line you drew. An example of a right triangle is shown below.

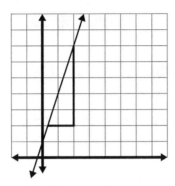

4. Label the vertices of the right triangle.

5. Find the length of the leg that is parallel to the *y*-axis (the vertical line segment) by subtracting the *y*-coordinates of its endpoints and finding the absolute value.

6. Find the length of the leg that is parallel to the *x*-axis (the horizontal line segment) by subtracting the *x*-coordinates of its endpoints and finding the absolute value.

7. Find the ratio of the length of the vertical line segment to the length of the horizontal line segment. This is the slope of the hypotenuse, which is a segment of the line you drew in Step 2.

8. Repeat Steps 3–7 two more times. Draw triangles that are similar to the first triangle you drew, using the same diagonal line as the hypotenuse. Remember that corresponding sides of similar triangles must have the same scale factor.

9. Answer the following questions:

   • What do you notice about the slope of the hypotenuse of each triangle?

   • What can you conclude about the slope of a line between any two points on the line?

STUDENT GUIDE FOR "I HAVE DERIVED IT"

This activity has two parts. First you will derive $y = mx$, which is the equation of a line through the origin. Next you will derive $y = mx + b$, which is the equation of a line intercepting the y-axis at b.

PART 1

1. Use your ruler to draw the x-axis and y-axis on your graph paper.

2. Draw a straight, non-vertical line that has a positive slope that intersects the origin.

3. Draw a right triangle so that one leg is on the x-axis and the other is parallel to the y-axis.

4. Label each vertex of the right triangle as shown.

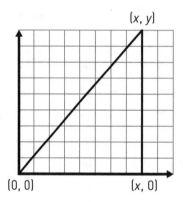

5. Find the length of the horizontal line segment.

6. Find the length of the vertical line segment.

7. Use the formula $m = \dfrac{\text{length of the vertical line segment}}{\text{length of the horizontal line segment}}$ and substitute the values you found in Step 5 and Step 6 in the correct places. This formula is equivalent to $m = \dfrac{\text{change in } y}{\text{change in } x}$.

8. Use the equation you wrote in Step 7 to solve for y.

PART 2

1. Use your ruler to draw the *x*- and *y*-axis on your graph paper.

2. Draw a straight, non-vertical line that has a positive slope that intersects the *y*-axis at *b*.

3. Draw a right triangle so that one leg is parallel to the *x*-axis and the other is parallel to the *y*-axis.

4. Label each vertex of the right triangle as shown.

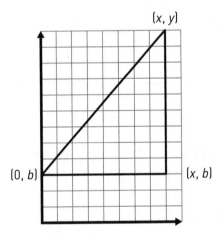

5. Find the length of the horizontal line segment.

6. Find the length of the vertical line segment.

7. Use the formula $m = \dfrac{\text{length of the vertical line segment}}{\text{length of the horizontal line segment}}$ and substitute the values you found in Step 5 and Step 6 in the correct places. This formula is equivalent to $m = \dfrac{\text{change in } y}{\text{change in } x}$.

8. Use the equation you wrote in Step 7 to solve for *y*.

Expressions and Equations: 8.EE.7

"Analyze and solve linear equations and pairs of simultaneous linear equations."

---

7. "Solve linear equations in one variable.

   a. "Give examples of linear equations in one variable with one solution, infinitely many solutions, or no solutions. Show which of these possibilities is the case by successfully transforming the given equation into simpler forms, until an equivalent equation of the form $x = a$, $a = a$, or $a = b$ results (where a and b are different numbers).

   b. "Solve linear equations with rational number coefficients, including equations whose solutions require expanding expressions using the distributive property and collecting like terms."

BACKGROUND

Equations in one variable can have one solution, infinitely many solutions, or no solutions. To solve for the value of the variable, students should follow the procedure for solving an equation that is summarized below:

- Use the distributive property $a(b + c) = ab + ac$ and $(b + c)a = ba + ca$ to eliminate parentheses, if necessary.

- Simplify one or both sides of the equation by collecting like terms, if necessary.

- Transform the equation into an equivalent equation by adding or subtracting the same number or expression to or from both sides of the equation, if necessary.

- Transform the equation into an equivalent equation by multiplying or dividing both sides of the equation by the same nonzero number.

The equations that follow have one solution.

$2(x - 4) + 2 = \frac{1}{2}x, x = 4$	$-x - 15 = 4x - 20, x = 1$	$\frac{2}{3}x + 7 = 11, x = 6$

The equations that follow have an infinite number of solutions. These equations can be transformed into equivalent equations, stating that two identical numbers are equal.

$3(x + 4) = 4x - x + 12$	$\frac{1}{2}x - 8 = 2\left(\frac{1}{4}x - 4\right)$	$7x + 4 = 3x + 4(x + 1)$

The equations that follow have no solutions. These equations can be transformed into equivalent equations, stating that two different numbers are equal.

$\frac{1}{4}x = \frac{3}{4}x - \frac{1}{2}x + 8$	$8x + 5 = 2(4x + 5)$	$\frac{1}{3}x + 5 = \frac{1}{3}(x - 15)$

 ## ACTIVITY: WHICH ONE DOESN'T BELONG?

Working in groups of three or four, students will write equations that either have one solution, an infinite number of solutions, or no solutions. They will arrange the equations in a grid so that two out of three equations in the same row have the same number of solutions. Another group of students will solve the equations and identify the equation in each row that does not belong with the other two equations in that row, based on the number of solutions.

PROCEDURE

1. Explain that each group must write nine equations. Three of the equations will have one solution, three will have an infinite number of solutions, and three will have no solutions.

2. Explain that after writing their equations, the group members should solve the equations and check that they do, in fact, have three equations with each type of solution.

3. Explain that students should sketch a table with three rows and three columns. They should place two equations that have the same type of solution in each of the three rows. They should then place one equation in each row that has a different type of solution than the other two equations in the row.

4. Offer the following example of a table that is constructed correctly. Ask your students to find which equation does not belong with the other two in its row.

$\frac{2}{3}x + 7 = 11$	$7x + 4 = 3x + 4(x + 1)$	$-x - 15 = 4x - 20$
$3(x + 4) = 4x - x + 12$	$\frac{1}{4}x = \frac{3}{4}x - \frac{1}{2}x + 8$	$\frac{1}{2}x - 8 = 2(\frac{1}{4}x - 4)$
$2(x - 4) + 2 = \frac{1}{2}x$	$8x + 5 = 2(4x + 5)$	$\frac{1}{3}x + 5 = \frac{1}{3}(x - 15)$

- Note that the second equation in Row 1 has an infinite number of solutions. It does not belong with the other two equations in that row, both of which have one solution, $x = 6$ and $x = 1$, respectively.

- In Row 2, the second equation does not belong because it has no solutions. The other two equations in Row 2 have an infinite number of solutions.

- In Row 3, the first equation does not belong because it has one solution, $x = 4$, and the other two equations have no solutions.

5. Instruct your students to make an answer key for their equations on another sheet of paper. Their answer key should identify which equation does not belong with the other two equations in each row, based on the number of solutions to the equations.

6. Remind students to include their names on both their grids and answer keys.

7. Instruct the groups to exchange their grids with that of another group. (Groups should retain their answer keys.)

8. Groups should find the equation in each row that does not belong.

CLOSURE

Check the results, using the answer keys of the groups. Ask your students which types of equations were most difficult to write and solve. Work as a class to solve any equations that students were unable to solve, or were unable to agree on the number of solutions.

Expressions and Equations: 8.EE.8

"Analyze and solve linear equations and pairs of simultaneous linear equations."

8. "Analyze and solve pairs of simultaneous linear equations.

   a. "Understand that solutions to a system of two linear equations in two variables correspond to points of intersection of their graphs, because points of intersection satisfy both equations simultaneously.

   b. "Solve systems of two linear equations in two variables algebraically, and estimate solutions by graphing the equations. Solve simple cases by inspection.

   c. "Solve real-world and mathematical problems leading to two linear equations in two variables."

BACKGROUND

A system of linear equations is two or more linear equations that have the same variables. A system of linear equations may have one solution, which is the point of intersection of two lines on a graph, no solution if the lines are parallel, or an infinite number of solutions if the lines coincide.

Students may solve a system of linear equations through a variety of methods, each of which is described below:

- *Graphing.* Students graph both equations in the same coordinate plane. The point of intersection is the solution to both equations. If the solutions are not integers, students will have to estimate the solutions.

- *Substitution.* Students solve one equation for one of the variables, and substitute this expression in the other equation to find the value of the other variable. They then substitute this value in either one of the original equations to find the value of the other variable.

- *Addition-or-subtraction.* Students will add or subtract the equations to obtain an equivalent equation that has only one variable, and then find the value of this variable. They will then substitute this value in one of the original equations and solve for the value of the other variable.

- *Multiplication with addition-or-subtraction.* Students will multiply one or both of the equations by the same non-zero number to obtain equivalent equations with the same or opposite coefficients of one of the variables. They will then follow the steps for using the addition-or-subtraction method.

ACTIVITY 1: WHAT'S THE POINT?

Students will virtually graph systems of linear equations and find the points of intersection.

MATERIALS

Computers with Internet access; digital projector.

PROCEDURE

1. Review the process of writing an equation in slope-intercept form, $y = mx + b$, because students must enter the equations they will graph in this form.

2. Go to http://my.hrw.com/math06_07/nsmedia/tools/Graph_Calculator/graphCalc.html to demonstrate the use of the virtual graphing calculator. Enter the equation $y_1 = 2x + 4$, then click on "Graph" to graph the line. Next click "Trace," and then click on the right arrow under "Zoom Out" to display the ordered pairs that lie on the line. Then enter $y_2 = -2x$ and follow the same procedure to graph the line and view the points on the line. The point of intersection is $(-1, 2)$. Note that the values $x = -1$ and $y = 2$ satisfy both equations. Also note that students can find the values that satisfy both equations by clicking on "Intersection," selecting the two equations, and clicking on "Find Intersection Point(s)."

3. Present each pair of linear equations that follow to your students. Instruct students to go to the Web site above to graph and find the values that satisfy each pair of equations simultaneously. Students should write each equation in slope-intercept form, if necessary, and enter the first equation as y_1 and enter the second equation as y_2. (The values of the points of the solutions are included below each system of equations.)

1) $y_1 = x - 1$ $y_2 = 2x + 1$ $(-2, -3)$	2) $y_1 = x - 2$ $y_2 = -x + 8$ $(5, 3)$	3) $y_1 = -x - 1$ $y_2 = 2x + 1$ $(-0.\overline{6}, -0.\overline{3})$
4) $y_1 = 3x$ $y_2 - 2x = 1$ $(1, 3)$	5) $y_1 - 4x = -2$ $y_2 = 2x + 1$ $(1.5, 4)$	6) $y_1 = 5x - 2$ $y_2 - 1 = 3x$ $(1.5, 5.5)$

4. Instruct your students to graph each pair of linear equations below that do not have one point of intersection. Instruct them to describe the graph and state if there are no solutions or an infinite number of solutions.

$y_1 - 2x = -1$ $y_2 = 2x + 1$ No solutions	$y_1 = x - 3$ $y_2 - x = -3$ Infinite number of solutions	$y_1 = 3x + 4$ $y_2 - 3x - 4 = 0$ Infinite number of solutions	$y_1 = 3x$ $y_2 - 3x = 5$ No solutions

Read the solutions to the systems of equations. Discuss that the graph of a system of linear equations has no solutions if the lines are parallel and has an infinite number of solutions if the lines coincide.

ACTIVITY 2: STUDENTS TEACHING

This is a two-day activity. Students will work in groups. Each group will create a short lesson that summarizes one way to solve systems of linear equations. Each group will create their lesson on the first day of the activity and present their lesson to the class on the second.

MATERIALS

Various items that are likely to be in your class, such as transparencies; markers for an overhead projector; graph paper; rulers; computer; digital projector.

PROCEDURE

1. Divide your students into four groups. Assign one of the methods used for solving systems of linear equations to each group: graphing, substitution, addition-or-subtraction, or multiplication with addition-or-subtraction. (If you have a very large class, you may divide students into more groups and allow two groups to do the same method.)

2. Explain that each group must create a five-minute lesson about their method, which they will then present to the class. They should include the following in their lesson:

   • A description of the method

   • Two examples of how to use the method

   • The benefit of using this method

   • Five problems that can be solved using the method, including a real-world application

   • An answer key

3. Recommend that the groups divide the work among their members. For example, some students develop a description of their method, some create examples for the use of the method, while others create problems to be solved and answer keys. Suggest that all group members brainstorm to find real-world applications.

4. Provide any materials the groups may need for their presentations.

5. Encourage your students to complete their lessons, and remind them that they will present their lessons during the next class. Suggest that each group appoint spokespersons to represent the group.

6. The next day, provide a few minutes for students to make any final adjustments to their lesson before beginning the presentations.

CLOSURE

Groups present their lessons to the class. At the end of each lesson, the class should solve the problems written by the group. Use the answer keys to confirm the correct solutions. Once all the presentations have been given, ask your students to explain which method, or methods, they found easiest to use.

Functions: 8.F.1

"Define, evaluate, and compare functions."

1. "Understand that a function is a rule that assigns to each input exactly one output. The graph of a function is the set of ordered pairs consisting of an input and the corresponding output."

BACKGROUND

A function is a rule that establishes a relationship between two quantities: an input and an output. For each input, there is exactly one output. Every input must be paired with exactly one output. If an input is paired with two or more different outputs, the relation is not a function.

Every function has a domain and a range. The domain of a function is the group of input values. The range of a function is the group of output values.

 ### ACTIVITY 1: USING A FUNCTION MACHINE

Students will use a function machine (a virtual manipulative) that assigns to each input exactly one output, find the output, and graph the function.

MATERIALS

Computers with Internet access; graph paper; rulers. Optional: digital projector.

PROCEDURE

1. Instruct your students to go to the Web site http://nlvm.usu.edu/en/nav/vlibrary.html. Direct them to click on "Algebra 6–8," scroll down, and click on "Function Machine."

2. Demonstrate the use of the virtual function machine and show your students what they are expected to do. Students will drag a number from the upper left-hand corner of the screen to the input of the function machine. Explain that the function machine uses a rule to produce an output value, and then places the value in a table, pairing the input with the output. Repeat the process of dragging a number into the function machine three more times.

3. Explain that students are to find the rule that the function machine is using to pair each input with one output. They can do this by considering the values of the input and output. Students should use this rule to complete a table. If students choose an output that is incorrect, students will receive an error message and should choose another output value.

4. Instruct your students to use the function machine, following Steps 2–3 on their own.

5. Students should copy the table and the function rule, then use graph paper and a ruler to graph the function.

6. When your students are finished, they can click on "New Function" for another function machine and table.

CLOSURE

Review your students' graphs. Discuss what is common to each graph. Students should conclude that for each value of x there is exactly one value for y.

 ## ACTIVITY 2: AND NOW FOR A SKIT, A SONG, A POEM, OR A...

This activity will require two class periods. Students will work in small groups to create a two- to three-minute presentation of a function, its domain and range. They will create their material on the first day and present it on the second.

MATERIALS

Student props (should you permit their use) should be simple, safe, nonbreakable, and easily obtained.

PROCEDURE

1. Divide your students into groups of three to four students.

2. Explain that they are to create a two- to three-minute presentation about a function. Offer suggestions of how they can present their function, such as:

   • A skit in which the resulting action is a function

   • A song or rap to express functions, domains, and ranges

   • A poem that defines functions, domains, and ranges

   Note that students may choose other means for their presentations; however, they should do so only with your approval.

3. Emphasize that the objective of the presentations is to explain what a function is and apply the definitions of domain and range. If necessary, review the following terms: function, input, output, domain, and range.

CLOSURE

Students give their presentations to the class. Discuss the activity, summarizing that a function is a rule that assigns to each input exactly one output.

Functions: 8.F.2

"Define, evaluate, and compare functions."

> 2. "Compare properties of two functions each represented in a different way (algebraically, graphically, numerically in tables or by verbal descriptions)."

BACKGROUND

A function is a rule that assigns to each input exactly one output. It can be represented by an equation, a graph, a table, or a verbal description. In each case, x is the input and y is the output.

For example, $y = 4x + 5$ is a function that multiplies the input by 4 and adds 5. It can also be represented as a graph of a straight line that has a slope of 4 and intercepts the y-axis at 5, or as a table that contains values such as $(-1, 1)$, $(0, 5)$, $(1, 9)$, and $(2, 13)$.

 ## ACTIVITY: WHAT'S MY FUNCTION?

Students will work with function cards for this activity. You will distribute 32 cards with a total of eight different functions, each function represented in four different ways. Each student in the class will receive at least one card that represents a function. Each student is to find three other students whose cards represent the same function, but in a different manner. Students will list the properties of their functions and compare these properties with other functions.

MATERIALS

Reproducibles, "Function Cards, I" and "Function Cards, II"; markers or chalk for writing on the board; rulers.

PREPARATION

Make one copy of "Function Cards, I" and "Function Cards, II." Cut out each card for a total of 32 cards. (Note that the cards on the reproducibles are arranged so that the cards in each row describe the same function. The originals serve as answer keys.)

PROCEDURE

1. Randomly distribute one card to each student. If you have fewer than 32 students in class, you may give some students two cards.

2. Explain that students are to circulate around the room to find three other students who have cards that describe the same function. These students should then sit together as

a group. (If a student received two cards, he should choose one of the cards and give the other one to a student of another group who has cards that describe the same function as this card. He should sit with the group that has cards that describe the function on the card he retained.) Once in their groups, students should verify that all of their cards describe the same function. If someone has a card that does not belong with the group, he or she should find the correct group.

3. Explain that students are to work in their groups and list the properties of their functions. They may include properties such as the slope, the y-intercept, and the quadrants that contain the line.

4. Explain that each group is to select one way to represent their function. One of the group's members will write this representation on the board. Underneath the representation, he or she will write the properties of the function.

5. After all of the groups have represented their function on the board, discuss the various properties of the functions, such as:

   • Which functions intersect the origin?

   • Which functions have the same rate of change?

   • Which function has the greatest rate of change?

   • Which function has the lowest rate of change?

CLOSURE

Instruct your students to write a description of the advantages and disadvantages of each representation of a function. They should include an example to support their reasoning.

$y = 3x$	x	y	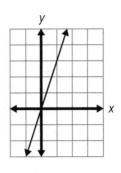	This function multiplies the input by 3.
	−1	−3		
	0	3		
	2	6		
	3	9		

$y = x$	x	y	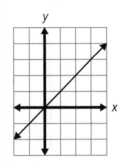	This function multiplies the input by 1.
	−2	−2		
	−1	−1		
	0	0		
	1	1		

$y = x + 1$	x	y	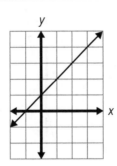	This function adds 1 to the input.
	−1	0		
	0	1		
	1	2		
	2	3		

$y = \frac{1}{2}x$	x	y	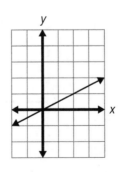	This function divides the input by 2.
	−2	−1		
	0	0		
	2	1		
	4	2		

$y = x - 1$	x	y		
	−2	−3	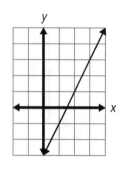	This function subtracts 1 from the input.
	−1	−2		
	0	−1		
	1	0		

$y = 2x - 3$	x	y		
	0	−3	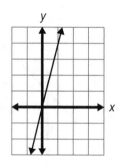	This function doubles the input, then subtracts 3 from it.
	1	−1		
	2	1		
	3	3		

$y = 4x$	x	y		
	−1	−4	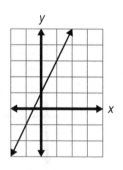	This function multiplies the input by 4.
	0	0		
	1	4		
	2	8		

$y = 2x + 1y$	x	y		
	−1	−1		This function doubles the input and adds 1 to it.
	0	1		
	1	3		
	2	5		

Functions: 8.F.3

"Define, evaluate, and compare functions."

> 3. "Interpret the equation $y = mx + b$ as defining a linear function, whose graph is a straight line; give examples of functions that are not linear."

BACKGROUND

All linear functions can be represented by the equation $y = mx + b$, where m is the slope of the line and b is the y-intercept. The point $(0, b)$ is the point where the line crosses the y-axis. The equation $y = mx + b$ contains two variables, x and y.

The graph of a linear function is a line. If a function is not linear, then its graph is not a line.

ACTIVITY: A FUNCTION SCAVENGER HUNT

For this activity, students will select equations from their math book, science text, other texts, or online sources to find functions and classify them as being linear or not linear. If the function is not linear, students will explain why it is not linear.

MATERIALS

Various texts and references; computers with Internet access.

PROCEDURE

1. Explain that a function is linear if it can be defined by the equation $y = mx + b$. Note that a linear function contains two variables and its graph is a line. Emphasize that functions that are not linear cannot be defined by the equation $y = mx + b$. The graph of a function that is not linear cannot be a line.

2. Instruct your students to find at least three examples of functions that are linear and three examples of functions that are not linear. Suggest that they search for examples in their math books, science texts, and other references, as well as online sources. They should explain how they determined if a function is linear or not linear.

3. Offer the following examples and explanations:

- $P = 4s$ is the formula for finding the perimeter of a square, given the length of a side, s. It is a linear function. The graph is a line that has a slope of 4 and intercepts the origin.

- $M = 0.17E$ is the formula for finding your weight on the Earth's moon, given your weight on the Earth, E. This is a linear function. The graph is a line that has a slope of 0.17 and intercepts the origin.

- $s = \sqrt{A}$ is the formula for finding the length of a side of a square, given the area, A. It is not a linear function. The graph contains the points $(0, 0)$, $(1, 1)$, and $(2, 4)$, which do not lie on a line.

- $V = \frac{4}{3}\pi r^3$ is the formula for finding the volume of a sphere, given the length of the radius, r. It is not a linear function. The graph contains the points $(0, 0)$, $\left(1, \frac{4}{3}\pi\right)$, and $\left(2, \frac{32}{3}\pi\right)$, which do not lie on a line.

CLOSURE

Ask your students to share their findings. Write some of their functions on the board, classifying them as being linear or not linear. Ask students to explain how they know whether or not a function is linear. Answers may vary, but students should note that the graph of a linear function is a non-vertical line and the graph of a function that is not linear is not a line.

Functions: 8.F.4

"Use functions to model relationships between quantities."

4. "Construct a function to model a linear relationship between two quantities. Determine the rate of change and initial value of the function from a description of a relationship or from two (x, y) values, including reading these from a table or from a graph. Interpret the rate of change and initial value of a linear function in terms of the situation it models, and in terms of its graph or a table of values."

BACKGROUND

Students can construct a function if they are given a relationship between two quantities, such as a verbal description of the relationship, two ordered pairs presented in a table, or values that they can obtain from a graph.

If they are given a verbal description, they can write an equation relating the two quantities.

If they are given two points, they can find the slope of the line that contains the points by using the formula $m = \dfrac{y_2 - y_1}{x_2 - x_1}$. They can substitute the slope and the x-coordinate and y-coordinate of an ordered pair into the equation $y = mx + b$ and solve for b. They can rewrite the equation $y = mx + b$ with the slope substituted for m and the *value* for b substituted for b.

If students are given a graph of a linear function, they may select two points on the line and follow the procedure above, or they may find the y-intercept and slope by analyzing the graph.

When students have defined a function in slope-intercept form, they can express the function by naming it with a letter, such as f. They can rewrite the equation in function notation by replacing y with $f(x)$. This is read as "the value of f at x" or "$f(x)$."

The initial value of the function is the value of the function when $x = 0$. Graphically, this is the point where the line intersects the y-axis, represented by b in the equation $y = mx + b$.

 ACTIVITY 1: FOUR IN A ROW

Students will work in pairs or groups of three to construct a function given a description, a table, or a graph. They will find the rate of change and initial value of the function.

MATERIALS

Scissors; glue sticks; reproducibles, "Constructing Functions, I," "Constructing Functions, II," and "Equation, Rate of Change, and Initial Value Cards."

PROCEDURE

1. Distribute one copy of each reproducible to each pair of students or group of students. Explain that the reproducibles "Constructing Functions, I" and "Constructing Functions, II" contain a total of eight functions that are described verbally, by a table, or by a graph. The reproducible "Equation, Rate of Change, and Initial Value Cards" contains 32 cards that students will cut out. All of the cards relate to one of the functions on the other two reproducibles. There are three different kinds of cards. Some cards contain an equation, some contain the rate of change, and others contain the initial value.

2. After cutting out the cards, instruct your students to glue the equation card, rate of change card, and initial value card that describes the function in the appropriate boxes on the sheets of "Constructing Functions, I" or "Constructing Functions, II" so that all four boxes in the row relate to the same function. Note that not all cards on the card sheet will be used.

3. The correct result will show eight functions (one function per row). Each row will contain a representation of a function, the equation, the rate of change, and the initial value.

CLOSURE

Provide the answers to your students.

ANSWERS

The answers are listed in the following order: equation, rate of change, and the initial value.

1) $y = x + 1$, 1, 1   **2)** $y = x + 3$, 1, 3   **3)** $y = 2x$, 2, 0   **4)** $y = -x + 1$, -1, 1

5) $y = -x - 7$, -1, -7   **6)** $y = 10x$, 10, 0   **7)** $y = -2x + 1$, -2, 1   **8)** $y = 6x + 23$, 6, 23

Ask your students to compare the equation, rate of change, and initial value of the functions with the representations that are provided. What similarities did they find? Their answers may vary, but students should understand that the rate of change is the same as the slope of the line, and that the initial value of the function is the same as the y-intercept of the graph.

 **ACTIVITY 2: EVERYDAY USES OF LINEAR FUNCTIONS**

Students will work in pairs or groups of three to find three examples of linear functions. They will then interpret the rate of change and the initial value in terms of the situation the functions model and in terms of their graphs.

MATERIALS

Graph paper; rulers; computers with Internet access.

PROCEDURE

1. Explain that a function can model a relation. A function is a rule that assigns to each input exactly one output. If the function models a linear relation, then the rate of change is constant.

2. Provide some examples of linear functions, such as:

- A teacher corrects 20 multiple-choice quizzes in 15 minutes. This can be modeled by a linear function that shows the number of quizzes she grades as a function of time in terms of hours. $y = mx$ or $f(x) = mx$. The rate of change is $\dfrac{20 \text{ quizzes}}{\frac{1}{4} \text{ hour}}$ or 80 quizzes per hour. The initial value (the value where $x = 0$) is 0. This means that when the time is 0, she is starting to grade the quizzes but has not yet graded any. This can be modeled by the function $f(x) = 80x$. The graph is a line with a slope of 80 that intersects the origin.

- A monthly cell phone plan that costs \$39.99 plus \$0.45 for each minute over 450 can be modeled by the function $f(x) = \$0.45x + \$39.99$. (As x represents the number of minutes greater than 450, you can set $x = 1$ to represent 451, $x = 2$ to represent 452 minutes, and so on.) The rate of change is \$0.45, which means that every minute in excess of 450 costs \$0.45 per minute, and the initial value is \$39.99. This can be expressed as when x is less than 1, no charge is made except for \$39.99. Graphically, the line $y = \$0.45x + \$39.99$ has a slope of 0.45 and intersects the y-axis at \$39.99.

3. Instruct your students to brainstorm to find real-world situations that can be modeled by linear functions, thinking of situations where the rate of change is constant. You might find it necessary to suggest some areas that can be modeled by linear functions, such as:

- Travel—averaging a rate for a given period of time

- Conversions—for example, Celsius to Fahrenheit, grams to ounces, and dollars to euros

- Admission fees and ticket prices

- Rental fees

- Taxi fares

4. Once your students have found a situation that can be modeled by a linear function, instruct them to do the following:

- Write the equation.

- Identify the rate of change and relate it to the situation.

- Identify the initial value and relate it to the situation.

- Draw the graph of the equation and interpret the slope and initial value.

CLOSURE

Provide time for students to share their results with the class. Ask your students what all of the functions have in common. Students should understand that every linear function has a constant rate of change.

#1

x	y
−1	0
0	1
1	2

#2

#3
The diameter of a circle is twice the radius.

#4

#5

x	y
−2	−5
1	−8
4	−11

#6
Mike saves $10.00 every week for a new iPad.

#7

#8

x	y
−4	−1
−3	5
−2	11

$y = -x - 7$	$y = 2x$	Rate of Change 1	Initial Value 0
$y = 3x$	$y = -x + 1$	Rate of Change 2	Initial Value 1
$y = 10x$	Rate of Change -1	Rate of Change -1	Initial Value 0
$y = x + 1$	Rate of Change -2	Rate of Change 1	Initial Value 23
$y = -2x + 1$	Rate of Change 1	Rate of Change 10	Initial Value 3
$y = 2x + 1$	Rate of Change 4	Initial Value 0	Initial Value 1
$y = x + 3$	Rate of Change 10	Initial Value 1	Initial Value 10
$y = 6x + 23$	Rate of Change 6	Initial Value -7	Initial Value 2

Functions: 8.F.5

"Use functions to model relationships between quantities."

5. "Describe qualitatively the functional relationship between two quantities by analyzing a graph (e.g., where the graph is increasing or decreasing, linear or nonlinear). Sketch a graph that exhibits the qualitative features of a function that has been described verbally."

BACKGROUND

A functional relationship between two quantities assigns to each input exactly one output. The graphs of a functional relationship may vary.

- The graph of a linear function, $f(x) = mx + b$, is a line that has a constant slope, m, and intersects the y-axis at b.

  - If $m = 0$, the graph of the function is a horizontal line.

  - If $m > 0$, the function is increasing. As the values of x increase, the corresponding y-values also increase.

  - If $m < 0$, the function is decreasing. As the values of x increase, the corresponding y-values decrease.

- The graph of a function that is nonlinear is a graph that is not a line. However, every value of x is paired with exactly one value of y. Some examples of functions that are nonlinear are the squaring function, $y = x^2$, the cubing function, $y = x^3$, and the square root function, $y = \sqrt{x}$.

 ACTIVITY: DESCRIBING A GRAPH

For this activity, students will first work individually and then in pairs or groups of three. Working individually, each student will write a verbal description of a graph, and then sketch the graph he described. Working with a partner, students will exchange the papers that contain the description of their graphs. They will then work individually again, this time to sketch the graph their partner described. Partners will then compare the graphs and determine if their sketches match the verbal descriptions and original sketches of each other's graph.

Graph paper; rulers.

1. Instruct your students to write a description of a graph. Suggest that they consider various possibilities for graphs. For example, Paulo might write "The graph is decreasing in quadrant II, and is increasing if *x* is greater than 0." Students should label their descriptions with their names. For example, "Paulo, Description 1." (Numbering the descriptions is helpful if students write additional descriptions.) Depending on your students, you may find it helpful to review that quadrants are numbered counterclockwise, starting with I in the upper right-hand quadrant.

2. After they have described their graphs, students should sketch their graphs on graph paper. They should label their graphs, for example, "Paulo, Graph 1." Paulo's graph might look like the following graph.

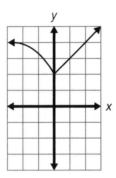

3. Instruct your students to exchange their descriptions with their partner. (For groups of three, students should exchange three ways.) Each student should try to draw her partner's graph, based on its description. They should label the graphs they draw, for example "Kelly's Drawing of Paulo's Description 1." Kelly's graph might look like the following graph.

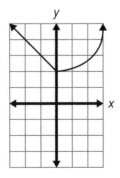

4. Ask your students to compare the graphs they drew with their partner's original graph. Do the graphs match the verbal descriptions? Are both graphs the same? In some cases, graphs may be different but may, in fact, match the verbal description.

5. Repeat the procedure to provide students with more practice in drawing graphs based on verbal descriptions.

CLOSURE

Discuss your students' descriptions and their graphs. Ask your students how they might write a description of a graph so that only one sketch fits the description. Although answers may vary, they should explain that the description must be specific. For example, instead of saying that a graph is increasing, you might include an equation as a part of the description.

Geometry: 8.G.1

"Understand congruence and similarity using physical models, transparencies, or geometry software."

> 1. "Verify experimentally the properties of rotations, reflections, and translations:
>
>   **a.** "Lines are taken to lines, and line segments to line segments of the same length.
>
>   **b.** "Angles are taken to angles of the same measure.
>
>   **c.** "Parallel lines are taken to parallel lines."

BACKGROUND

Rotations, reflections, and translations are rigid transformations. As an original figure, called the pre-image, is rotated, reflected, or translated, a congruent figure, called the image, takes a different position or has a different orientation. The image has the same size and shape as the pre-image.

An explanation of each transformation follows:

- *Rotation:* a figure is rotated around a fixed point in either a clockwise or counterclockwise direction.

- *Reflection:* a figure is reflected, or flipped, over a line. Each figure is a mirror image of the other.

- *Translation:* a figure is moved a specific number of units up, down, left, or right.

ACTIVITY: TRANSFORMATIONS

Students will draw intersecting lines, line segments, angles, and parallel lines. They will rotate, reflect, and translate these figures to determine that the lengths, angles measures, and distances between parallel lines are the same.

MATERIALS

1 transparency sheet for every four students; rulers; nonpermanent markers; tissues or paper towels; 2-page reproducible, "Using Transformations."

PREPARATION

Cut each transparency into 4 pieces measuring about 4″ × 5″. (Each student will receive one piece.)

PROCEDURE

1. Explain that students will draw intersecting lines, line segments, angles, and parallel lines on plain paper. These figures are called pre-images. Next, students will place a transparency over each figure and trace the figure. Once the figures are traced, they can be transformed. Note that students will then be asked to compare the images with the pre-images.

2. Review the directions on the reproducible, "Using Transformations," with your students. If necessary, discuss the vocabulary, which will help students complete the activity.

3. Emphasize that students should follow the directions on the reproducible closely.

CLOSURE

Ask your students to explain how they know that when a figure is rotated, reflected, or translated, lines are taken to lines, line segments are taken to line segments of the same measure, and angles are taken to angles of the same measure. Although answers will likely vary, a possible answer is that the image has the same size and shape as the pre-image. Therefore, its attributes must be the same.

USING TRANSFORMATIONS

You will draw various geometric figures such as lines, line segments, angles, and parallel lines. You will then rotate, reflect, and translate these figures, comparing their transformations with the original figures.

Words to know:

- Image: A geometric figure after it has been transformed.

- Pre-image: An original geometric figure that has not been moved or changed.

- Reflection: A flip of a geometric figure over a line of reflection to create a mirror image.

- Rotation: A turn of a geometric figure about a fixed point without reflection.

- Transformation: The change in the position of a geometric figure, the pre-image, that produces a new figure called the image.

- Translation: A movement of a geometric figure to a new position by sliding along a straight line.

Follow these steps:

1. Use a ruler to draw two nonparallel lines, called a pre-image, on plain paper.

2. Place a transparency over the lines.

3. Use nonpermanent markers to trace the lines onto the transparency.

4. Rotate the image on the transparency by placing the tip of your pencil on the transparency and turning (rotating) the transparency around the point. Note that the new figure, called the image, has the same size and shape as the pre-image, but it is in a different position.

5. Place the transparency over the pre-image (the two nonparallel lines drawn in Step 1).

(continued)

6. Flip (reflect) the image on the transparency. This is similar to turning a page in a book. The image on the transparency is a reflection of the pre-image. It is a mirror image of the lines you drew. The line of reflection is the edge of the transparency. Note that the image has the same size and shape as the pre-image, but it is in a different position.

7. Place the transparency over the pre-image (the two nonparallel lines drawn in Step 1).

8. Translate the transparency by moving it up or down, to the left, or to the right. You may move it in combinations; for example, up and to the left, or down and to the right. Note that the image has the same size and shape as the pre-image, but it is in a different position.

9. Clean your transparency with tissue paper or a paper towel.

10. Follow Steps 1–9 three more times. Draw line segments, angles, and then parallel lines.

Geometry: 8.G.2

"Understand congruence and similarity using physical models, transparencies, or geometry software."

2. "Understand that a two-dimensional object is congruent to another if the second can be obtained from the first by a series of rotations, reflections, and translations; given two congruent figures, describe a sequence that exhibits the congruence between them."

BACKGROUND

Rotations, reflections, and translations are transformations that preserve lengths and distances. Each image formed by one or a combination of these transformations is congruent.

 ACTIVITY 1: FIND THE IMAGE

Working in pairs or groups of three, students will investigate transformations, first manually and then virtually. They will select a square, parallelogram, or triangle. They will then select an angle of rotation, line of reflection, and/or the number of units to translate a figure. After transforming their figure, they will draw its image, then verify their results virtually using the transmographer.

MATERIALS

Computers with Internet access; digital projector; graph paper; rulers; protractors; 2-page reproducible, "Student Guide for Transforming Figures." Optional: scissors.

PROCEDURE

1. Distribute a copy of the reproducible, "Student Guide for Transforming Figures," to each student and review the information. Emphasize that they should follow the guidelines as they complete the activity. (You may want to inform students to retain the reproducible upon conclusion of this activity. This reproducible will be used in subsequent activities.)

2. Instruct your students to go to the Web site http://www.shodor.org/interactive/activities /transmographer/.

3. Demonstrate the use of the transmographer for your students.

- Select a polygon by clicking on a new square, a new parallelogram, or a new triangle.

- Note that once the polygon is selected, it is graphed in the coordinate plane. Each vertex is labeled and has a different color.

- Explain that to rotate a polygon, students must select a point and the number of degrees. All rotations are counterclockwise on the transmographer. Once students click on "Rotate," the image is graphed. The vertices of the image are color-coded, corresponding to the vertices of the pre-image. The vertices are also labeled.

- Explain that to reflect a polygon, students must specify the line of reflection. Once they click on "Reflect," the image is graphed. The vertices of the image are color-coded, corresponding to the vertices of the pre-image. The vertices are also labeled.

- Explain that to translate a polygon, students must specify the number of units on the x-axis, the y-axis, or both axes. Once students click on "Translate," the image is graphed. The vertices of the image are color-coded, corresponding to the vertices of the pre-image. The vertices are also labeled.

4. Provide time for students to experiment and learn how to use the transmographer.

5. After your students have gained competency with the use of the transmographer, explain that they are to select a figure on the screen by clicking on a new square, a new parallelogram, or a new triangle. For example, they might choose a square. Note that as students use the transmographer to select the square, the square will be graphed on the screen.

6. Instruct your students to use graph paper and graph a square that has the same vertices as the square on the screen. Next, they should reflect the square across a line, for example, $x = 2$, on their graph paper. To verify the accuracy of the graph they drew, they must enter $x = 2$ on the screen and then click on "Reflect." Students should compare their hand-drawn image with the image on the screen. Note that this process allows students to check the accuracy of hand-drawn images. Encourage your students to rotate and translate the square. Allow time for students to select other polygons and transform them.

7. If students have difficulty visualizing and drawing the transformations, provide scissors and protractors so that they can cut out the pre-images on their graph paper and physically translate them.

CLOSURE

Ask your students which transformations were easiest to perform and why. (Answers may vary.) Also ask how students know that an image is congruent to its pre-image. They should understand that the size and shape of the image has not changed. Only its position has changed.

ACTIVITY 2: FROM HERE TO THERE

Working individually first, then in pairs or groups of three, students will describe a sequence of transformations to "move" a pre-image to an image.

MATERIALS

Graph paper; rulers; protractors; a set of pattern blocks for each student; reproducible, "Student Guide for Transforming Figures."

PROCEDURE

1. Distribute a copy of the reproducible and a set of pattern blocks to each student. For the first part of this activity, students work individually. Review the information on the reproducible, and, if necessary, explain the use of the pattern blocks.

2. Instruct your students to trace a polygon (the pre-image) from the pattern block set onto graph paper. They should then rotate, reflect, and/or translate the figure, tracing the image as they proceed. As they trace the image, they should write down the steps for the transformation. Remind them to follow the guidelines on the reproducible when transforming figures.

3. Instruct students to trace the pre-image on another sheet of graph paper in its original position, and then exchange the pre-image with that of another student. Students will now work in pairs or groups of three.

4. One student will read the written instructions for her transformation to her partner (or partners), who will follow the directions and draw the image. They will then compare the pre-image with the image. If the pre-image and image do not match, they should work together and review each step to identify any errors.

5. The other student will then read her written instructions for her transformation to her partner, who will follow the directions and draw the image. Students should then compare this pre-image with its image and resolve any discrepancies.

CLOSURE

Discuss the activity with the class, explaining the importance of accuracy in describing transformations.

STUDENT GUIDE FOR TRANSFORMING FIGURES

Figures can be transformed by rotations, reflections, and translations, or a combination of these actions. In a transformation, lengths, distances, and any angles of the figures remain constant.

A **rotation** is a transformation in which a pre-image is turned (either clockwise or counterclockwise) about a specific point called the center of rotation. To rotate a polygon a specified number of degrees, do the following:

1. Determine the center of rotation and the direction of rotation.

2. Draw a ray from the center of rotation to the vertex of the polygon.

3. Use a protractor to draw the other ray of the angle of rotation.

4. Mark the image of the vertex.

5. Continue Steps 2–4 until all vertices have been rotated.

6. Draw the image by drawing line segments connecting adjacent vertices.

The following example shows a triangle rotated 90° counterclockwise about the origin.

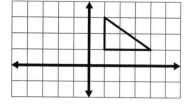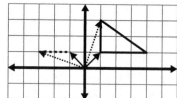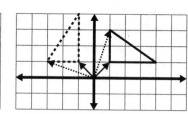

A **reflection** is a transformation in which a figure is flipped over a line called a line of reflection. This line can be thought of as a mirror in which the image is a reflection of the pre-image. To reflect a polygon, do the following:

1. Determine the line of reflection.

2. Find the distance from the vertex of the polygon to the line of reflection.

3. Mark the image of the vertex the same distance from the line of reflection, but in the opposite direction.

4. Continue Steps 2 and 3 until all vertices have been reflected.

5. Draw the image by drawing line segments connecting adjacent vertices.

The following example shows a triangle reflected over the *y*-axis.

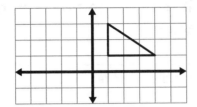 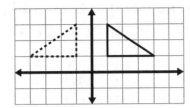

A **translation** is a transformation in which a figure is moved horizontally, vertically, or in a combination of these directions. To translate a polygon, do the following:

1. Determine whether the movement will be horizontal, vertical, or a combination of horizontal and vertical movements.

2. Find the distance from the vertex of the polygon to the new position.

3. Mark the image on a vertex.

4. Continue Steps 2 and 3 until all vertices have been translated.

5. Draw the image by drawing line segments connecting adjacent vertices.

The following example shows a triangle translated 4 units to the left.

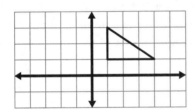 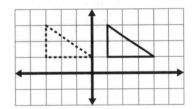

Geometry: 8.G.3

"Understand congruence and similarity using physical models, transparencies, or geometry software."

3. "Describe the effect of dilations, translations, rotations, and reflections on two-dimensional figures using coordinates."

BACKGROUND

Rotations, reflections, and translations produce an image that is congruent to the pre-image. A dilation produces an image that is similar to the pre-image. It has the same shape but not necessarily the same size. Corresponding angles are congruent and corresponding sides are k times as long as the pre-image, where k is the scale factor.

- If $k > 1$, the dilation is an enlargement.
- If $0 < k < 1$, the dilation is a reduction.

 ### ACTIVITY: I'VE SCRAMBLED MY NOTES

Working in pairs or groups of three, students will be given the coordinates of a pre-image. They will match a transformation of the pre-image with an explanation about how to find the coordinates of its image. They will then match the coordinates of the image with the correct transformation.

MATERIALS

Scissors; glue sticks; graph paper; rulers; reproducibles, "Student Guide for Dilating Figures" and 2-page "The Mixed-Up Notes." (You may also find it helpful to use the reproducible, "Student Guide for Transforming Figures," which is included with Geometry: 8.G.2.)

PROCEDURE

1. Explain that the transformations in this activity include rotations, reflections, translations, and dilations. Remind your students that images formed by rotating, reflecting, or translating produce an image that is congruent to the pre-image. (Congruent figures have the same shape and size.) A dilation produces an image that is similar to the pre-image. (Similar figures have the same shape but are not necessarily the same size.)

2. Distribute copies of the reproducibles "Student Guide for Dilating Figures" and "The Mixed-Up Notes" to each student. Review the information on the reproducibles, and read the premise on "The Mixed-Up Notes" with your students, emphasizing that the pre-image for every transformation has coordinates: A (-3, 4), B (2, 4), and C (-2, 1). Because students will also rotate, reflect, and translate figures, you may find it helpful to distribute copies of the reproducible "Student Guide for Transforming Figures." Students may also find it helpful to sketch figures on graph paper.

3. Encourage your students to refer to the reproducibles as they complete the activity. If they have trouble finding the coordinates of the images, suggest that they sketch the pre-images and the images on graph paper.

CLOSURE

Correct the answers of your students. If some students have difficulty finding images, demonstrate the steps of the transformation.

ANSWERS

Answers are arranged from top to bottom.

Column II: 7, 4, 1, 9, 5, 8, 3, 10, 2, 6
Column III: 13, 19, 17, 20, 14, 11, 16, 18, 15, 12

STUDENT GUIDE FOR DILATING FIGURES

Dilations are transformations in which a pre-image is either enlarged or reduced. The image produced is similar to the pre-image. It has the same shape, but it is not necessarily the same size.

All dilations require a center of dilation and a scale factor. The center of dilation is a point from which every point on the pre-image will be moved toward or away from. The scale factor is the ratio of the corresponding sides of the image to the pre-image. If the scale factor is larger than 1, the image is an enlargement. If the scale factor is between 0 and 1, the image is a reduction.

To dilate a figure, do the following:

1. Determine the center of dilation and the scale factor.

2. Draw a ray from the center of dilation to each vertex of the figure.

3. Measure the distance from the center of dilation to a vertex of the figure.

4. Multiply this distance by the scale factor.

5. Mark this distance on the ray.

6. Repeat Steps 3–5 until each of the distances from the center of dilation to the vertices of the image have been marked.

7. Draw the image by drawing line segments connecting the vertices.

The following example shows a triangle that is dilated, using the origin as the center of dilation. The scale factor is equal to 3.

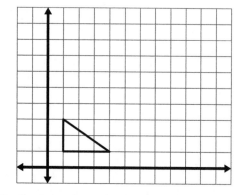 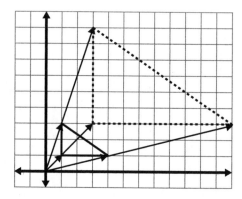

To find the length of a segment of the image that has been dilated, multiply the corresponding length of its pre-image by the scale factor.

To find the coordinates of the image, multiply the corresponding *x*- and *y*-values by the scale factor.

THE MIXED-UP NOTES

Sara is making a PowerPoint presentation, summarizing the transformations she learned about this year. Every transformation is listed below in Column I. In Column II, she described how to find the coordinates of the image formed by the transformation listed in Column I. In Column III, she listed the coordinates of the image that was formed using the transformation in Column I.

All transformations use the pre-image of a triangle with coordinates: A (−3, 4), B (2, 4), and C (−2, 1).

Sara has a problem, though. She mixed up her notes for Columns II and III. She tried to correct her notes, but she made errors.

Help Sara correct her notes. Cut out the notes in Column II and Column III. Glue the transformations in Column I (in their original order) on two blank sheets of paper, along the left margin. Glue notes numbered 1–10 in the correct order in Column II. Then glue notes numbered 11–20 in Column III so that the coordinates and descriptions match the transformation described in Column I.

Column I	Column II	Column III
A translation parallel to the x-axis (for example, 2 units to the left)	**1.** Add the number of units that are moved parallel to the x-axis to the x-coordinates of the pre-image. Add the number of units that are moved parallel to the y-axis to the y-coordinates of the pre-image.	**11.** (4, −3), (4, 2), (1, −2)
A translation parallel to the y-axis (for example, 3 units up)	**2.** The x-coordinates of the image have the same values of the y-coordinates of the pre-image. The y-coordinates of the image are the opposite of the x-coordinates of the pre-image.	**12.** (−6, 8), (4, 8), (−4, 2)
A translation that is a combination of translations parallel to the x-axis and parallel to the y-axis (for example, 1 unit to the right and 2 units down)	**3.** The x-coordinates of the image are the opposite of the y-coordinates of the pre-image. The y-coordinates of the image are the same as the x-coordinates of the pre-image.	**13.** (−5, 4), (0, 4), (−4, 1)

(continued)

A reflection in the *y*-axis	**4.** Add the number of units to the *y*-coordinates of the pre-image. The values of the *x*-coordinates remain the same.	**14.** (−3, −4), (2, −4), (−2, −1)
A reflection in the *x*-axis	**5.** The *y*-coordinates of the image are the opposite of the *y*-coordinates of the pre-image. The *x*-coordinates are the same.	**15.** (4, 3), (4, −2), (1, 2)
A reflection in the line *y* = *x*	**6.** The *x*-coordinates of the image are double the *x*-coordinates of the pre-image. The *y*-coordinates of the image are double the *y*-coordinates of the pre-image.	**16.** (−4, −3), (−4, 2), (−1, −2)
A rotation of 90°counterclock-wise around the origin	**7.** Add the number of units to the *x*-coordinates of the pre-image. The values of the *y*-coordinates remain the same.	**17.** (−2, 2), (3, 2), (−1, −1)
A rotation of 180° around the origin	**8.** The *x*-coordinates of the image is the same value as the *y*-coordinates of the pre-image. The *y*-coordinates of the image is the same as the *x*-coordinates of the pre-image.	**18.** (3, −4), (−2, −4), (2, −1)
A rotation of 270° counterclockwise around the origin (or 90° clockwise)	**9.** The *x*-coordinates of the image are the opposite of the *x*-coordinates of the pre-image. The *y*-coordinates are the same.	**19.** (−3, 7), (2, 7), (−2, 4)
A dilation about the origin (For example, the scale factor is 2.)	**10.** The *x*-coordinates of the image are the opposite of the *x*-coordinates of the pre-image. The *y*-coordinates of the image are the opposite of the *y*-coordinates of the pre-image.	**20.** (3, 4), (−2, 4), (2, 1)

Geometry: 8.G.4

"Understand congruence and similarity using physical models, transparencies, or geometry software."

4. "Understand that a two-dimensional figure is similar to another if the second can be obtained from the first by a sequence of rotations, reflections, translations, and dilations; given two similar two-dimensional figures, describe a sequence that exhibits the similarity between them."

BACKGROUND

Rotations, reflections, and translations of a pre-image produce an image that is congruent to the pre-image. A dilation produces an image that is similar to the pre-image. If the scale factor is greater than 1, the image is an enlargement. If the scale factor is between 0 and 1, the image is a reduction of the pre-image. A sequence of rotations, reflections, transformations, and dilations will produce an image that is similar to the pre-image.

 ### ACTIVITY: I FOUND THE IMAGE

Working in groups of three or four, students will determine the image of a rectangle that has been rotated, reflected, translated, and dilated.

MATERIALS

Graph paper; 2-page rulers; reproducible, "Group Tasks for Finding the Image." Optional: scissors; 2-page reproducible, "Student Guide for Transforming Figures" which is included with Geometry: 8.G.2 and reproducible, "Student Guide for Dilating Figures," included with Geometry 8.G.3.

PREPARATION

Make enough copies of the reproducible, "Group Tasks for Finding the Image," so that every student has a copy of his or her group's tasks for this activity. Cut out the directions for each group.

PROCEDURE

1. Designate each group with a number from 1 to 5. Distribute a copy of the group's tasks to each member of each group. (For large classes, two groups might work on the same task. Simply make additional copies of the tasks and label the additional group tasks as 1a, 2a, etc.)

2. Explain that each group has tasks to complete. The members of each group should work together to complete the group's task, which is to graph a pre-image and perform a variety of specific transformations. Note that there are six steps, each of which results in a new image. Students should record the vertices of the image after each transformation. Note that accuracy is critical, because each image becomes the pre-image in the next step.

3. You may find it helpful to distribute copies of reproducible "Student Guide for Transforming Figures," and reproducible "Student Guide for Dilating a Figure," should students need guidance for performing transformations.

4. Monitor students' work. Once students have finished, ask a student in each group to read the coordinates of the last image the group found. This will enable you to check if their work is correct. If they made a mistake with any of the transformations, the coordinates of the final image will not be correct. (The correct coordinates are listed under Closure.) If students' final images are incorrect, ask them to recheck their work, focusing on the vertices of the images of the transformations to find their mistake.

CLOSURE

Discuss the activity. Explain that each group had to graph the same pre-image, with the same coordinates, and the same set of instructions, but in a different order. Each group, therefore, found a different image at Step 6. Ask your students what this implies about the order of transformations. They should realize that the order of performing transformations matters.

ANSWERS

Following are the answers to the group tasks. (Note: For each group, the first set of coordinates are given on the group's task sheet.)

Group 1. 2: (1, 4), (2, 4), (2, 2), (1, 2). 3: (1, −4), (2, −4), (2, −2), (1, −2). 4: (−1, 4), (−2, 4), (−2, 2), (−1, 2). 5: (−1, 7), (−2, 7), (−2, 5), (−1, 5). 6: (−3, 7), (−4, 7), (−4, 5), (−3, 5)

Group 2. 2: (−2, −8), (−4, −8), (−4, −4), (−2, −4). 3: (−4, −8), (−6, −8), (−6, −4), (−4, −4). 4: (−2, −4), (−3, −4), (−3, −2), (−2, −2). 5: (−2, 4), (−3, 4), (−3, 2), (−2, 2). 6: (−2, 7), (−3, 7), (−3, 5), (−2, 5)

Group 3. 2: (1, 4), (2, 4), (2, 2), (1, 2). 3: (−1, 4), (0, 4), (0, 2), (−1, 2). 4: (1, −4), (0, −4), (0, −2), (1, −2). 5: (1, 4), (0, 4), (0, 2), (1, 2). 6: (1, 7), (0, 7), (0, 5), (1, 5)

Group 4. 2: (1, 4), (2, 4), (2, 2), (1, 2). 3: (−1, 4), (0, 4), (0, 2), (−1, 2). 4: (−1, −4), (0, −4), (0, −2), (−1, −2). 5: (−1, −1), (0, −1), (0, 1), (−1, 1). 6: (1, 1), (0, 1), (0, −1), (1, −1)

Group 5. 2: (2, −8), (4, −8), (4, −4), (2, −4). 3: (1, −4), (2, −4), (2, −2), (1, −2). 4: (1, −1), (2, −1), (2, 1), (1, 1). 5: (−1, 1), (−2, 1), (−2, −1), (−1, −1). 6: (−3, 1), (−4, 1), (−4, −1), (−3, −1)

GROUP TASKS FOR FINDING THE IMAGE

--- ✂ ---

GROUP 1 TASKS

1. Graph a rectangle by graphing points A (2, 8), B (4, 8), C (4, 4), and D (2, 4).

2. Dilate the figure with respect to the origin. The scale factor is $\frac{1}{2}$. Record the vertices of the image.

3. Reflect the image in $y = 0$. Record the vertices of the image.

4. Rotate the image 180° around the origin. Record the vertices of the image.

5. Translate the figure up 3 units. Record the vertices of the image.

6. Translate the figure 2 units to the left. Record the vertices of the image.

--- ✂ ---

GROUP 2 TASKS

1. Graph a rectangle by graphing points A (2, 8), B (4, 8), C (4, 4), and D (2, 4).

2. Rotate the image 180° around the origin. Record the vertices of the image.

3. Translate the figure 2 units to the left. Record the vertices of the image.

4. Dilate the figure with respect to the origin. The scale factor is $\frac{1}{2}$. Record the vertices of the image.

5. Reflect the image in $y = 0$. Record the vertices of the image.

6. Translate the figure up 3 units. Record the vertices of the image.

--- ✂ ---

GROUP 3 TASKS

1. Graph a rectangle by graphing points A (2, 8), B (4, 8), C (4, 4), and D (2, 4).

2. Dilate the figure with respect to the origin. The scale factor is $\frac{1}{2}$. Record the vertices of the image.

3. Translate the figure 2 units to the left. Record the vertices of the image.

4. Rotate the image 180° around the origin. Record the vertices of the image.

5. Reflect the image in $y = 0$. Record the vertices of the image.

6. Translate the figure up 3 units. Record the vertices of the image.

--- ✂ ---

(continued)

GROUP TASKS FOR FINDING THE IMAGE (continued)

- ✂ - - -

GROUP 4 TASKS

1. Graph a rectangle by graphing points A (2, 8), B (4, 8), C (4, 4), and D (2, 4).

2. Dilate the figure with respect to the origin. The scale factor is $\frac{1}{2}$. Record the vertices of the image.

3. Translate the figure 2 units to the left. Record the vertices of the image.

4. Reflect the image in $y = 0$. Record the vertices of the image.

5. Translate the figure up 3 units. Record the vertices of the image.

6. Rotate the image 180° around the origin. Record the vertices of the image.

- ✂ - - -

GROUP 5 TASKS

1. Graph a rectangle by graphing points A (2, 8), B (4, 8), C (4, 4), and D (2, 4).

2. Reflect the image in $y = 0$. Record the vertices of the image.

3. Dilate the figure with respect to the origin. The scale factor is $\frac{1}{2}$. Record the vertices of the image.

4. Translate the figure up 3 units. Record the vertices of the image.

5. Rotate the image 180° around the origin. Record the vertices of the image.

6. Translate the figure 2 units to the left. Record the vertices of the image.

- ✂ - - -

Geometry: 8.G.5

"Understand congruence and similarity using physical models, transparencies, or geometry software."

> 5. "Use informal arguments to establish facts about the angle sum and exterior angle of triangles, about the angles created when parallel lines are cut by a transversal, and the angle-angle criterion for similarity of triangles."

BACKGROUND

Listed below are facts about angles of a triangle, angles created when parallel lines are cut by a transversal, and criterion for similarity of triangles:

- The sum of the interior angles of a triangle is $180°$.

- The sum of the exterior angles of a triangle is $360°$.

- When two parallel lines are cut by a transversal, five pairs of angles are formed:

  - *Alternate interior angles:* a pair of non-adjacent angles that lie between the parallel lines on opposite sides of the transversal.

  - *Alternate exterior angles:* a pair of non-adjacent angles that lie outside the parallel lines on opposite sides of the transversal.

  - *Same side interior angles (also called consecutive interior angles):* a pair of angles that lie between the parallel lines on the same side of the transversal.

  - *Same side exterior angles:* a pair of angles that lie outside the parallel lines on the same side of the transversal.

  - *Corresponding angles:* a pair of angles that are in the same position with respect to the transversal and parallel lines.

- When two lines intersect, two pairs of angles are formed:

  - *Vertical angles:* two angles whose sides form two pairs of opposite rays.

  - *Adjacent angles:* two angles that share a common vertex and a common side but have no common interior points.

- The following pairs of angles are congruent:

  - Alternate interior angles

  - Alternate exterior angles

  - Corresponding angles

  - Vertical angles

- The following pairs of angles are supplementary:
  - Same side interior angles
  - Same side exterior angles
  - Adjacent angles
- The angle-angle similarity postulate states that if two angles of one triangle are congruent to two angles of another triangle, then the triangles are similar.

 ACTIVITY 1: FINDING THE SUM OF THE INTERIOR ANGLES OF A TRIANGLE

This is a two-part activity. For the first part of the activity, students will work at a Web site where they will virtually demonstrate that the sum of the interior angles of a triangle equals 180° by dragging a vertex of a triangle. They will note how this affects the other measures of the angles of the triangle. For the second part of the activity, students will draw and then cut a triangle and rearrange the angles to form a line, showing that the sum of the angles in a triangle equals 180°.

MATERIALS

Computers with Internet access; graph paper; rulers; scissors.

PROCEDURE

1. Explain to your students that they will find the sum of the angles of a triangle.

2. Instruct them to go to the Web site www.mathwarehouse.com/geometry/triangles /interactive-triangle.htm.

3. Instruct them to drag a vertex of the triangle and note the measure of each angle that is displayed. Next they should find the sum of the angle measures. Note that the sum is always 180°.

4. Allow time for your students to continue to investigate angle measures by dragging the vertices of the triangle in various directions. They should verify the sum of the measures of the angles.

5. Explain that students will now use a model to find the sum of the angles of a triangle.

6. Instruct them to draw a large triangle on graph paper and cut it out.

7. They should label the vertices a, b, and c inside the angles. Then they should cut the triangle so that the original triangle is now divided into three pieces, each piece with an angle labeled a, b, or c.

8. Instruct your students to rearrange the vertices so that a common point is the vertex of all three angles and the sides touch but do not overlap. Students will likely find different arrangements, but all arrangements will form a line. Ask your students what

this proves about the sum of the measures of the angles of a triangle. Students should conclude that the angles arranged in this manner form a line (also known as a straight angle) that contains 180°.

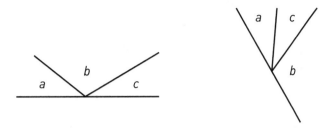

Examples of Two Possible Angle Arrangements

CLOSURE

Ask your students to write an explanation of how they know the sum of the angles in a triangle equals 180°. Students' explanations should note that the angles form a line that contains 180°.

 ## ACTIVITY 2: FINDING THE SUM OF THE EXTERIOR ANGLES OF A TRIANGLE

For this activity, students will work at a Web site and virtually create an exterior angle of a triangle, apply the exterior angle theorem, and find the sum of the exterior angles of a triangle algebraically.

MATERIALS

Computers with Internet access; reproducible, "Finding the Sum of the Exterior Angles of a Triangle."

PROCEDURE

1. Explain that students will visit a Web site (noted on the reproducible) and explore facts about the exterior angles of a triangle. They will virtually create an exterior angle of a triangle; examine the relationship between the exterior angle and the sum of its remote interior angles; and use algebra to determine the sum of the angles of a triangle.

2. Distribute copies of the reproducible, and note that students are to follow the guidelines to complete the activity.

ACTIVITY 3: ANGLES, PARALLEL LINES, AND TRANSVERSALS

Working at a Web site, students will identify the types of angles formed by two parallel lines that are cut by a transversal. They will also draw parallel lines that are cut by a transversal, trace their drawings, and transform their images to determine relationships between angles.

MATERIALS

Computers with Internet access; rulers; graph paper (two sheets per student); tracing paper (two sheets per student).

PROCEDURE

1. Explain that when two parallel lines are cut by a transversal, several pairs of angles are formed: alternate interior angles, alternate exterior angles, same side interior angles, same side exterior angles, and corresponding angles. Note that vertical angles and adjacent angles are formed when any two lines intersect.

2. Instruct your students to go to the Web site http://www.shodor.org/interactivate /activities/Angles. Explain that they will be given a diagram and they will identify the types of angles formed when two parallel lines are cut by a transversal.

3. Explain that students will receive immediate feedback by clicking on "Check Answers." They will have the opportunity to change their answers and correct them. They can also click on "New Angles" to generate other diagrams.

4. After your students have become familiar with these types of angles presented on the Web site, they will be ready to make and test conjectures about the types of angles formed when two parallel lines are cut by a transversal.

5. Instruct your students to draw two parallel lines on graph paper. Next, they should draw a transversal that intersects both parallel lines. They should label the angles $\angle a$ through $\angle h$ as noted on the Web site and on the following diagram.

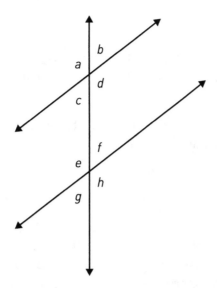

6. Next, students should use tracing paper to trace the lines and angle labels they drew. Instruct them to move the tracing paper by translating it or rotating it to find congruent and supplementary angles. For example, ∠e is congruent to ∠a and ∠e is supplementary to ∠c.

7. Instruct your students to record all congruent angles and all supplementary angles.

8. To provide students with more practice, suggest that they draw another pair of parallel lines, cut the lines with a transversal, and repeat the process.

CLOSURE

Ask your students what types of angles are congruent and what types of angles are supplementary. They should realize that alternate interior angles, alternate exterior angles, corresponding angles, and vertical angles are congruent. Same side interior and same side exterior and adjacent angles are supplementary.

ACTIVITY 4: ANGLE-ANGLE SIMILARITY

For this activity, students will draw two triangles to demonstrate the angle-angle similarity postulate.

MATERIALS

Unlined paper; rulers; protractors; reproducible, "Drawing Similar Triangles."

1. Explain to your students that they will draw two triangles to find the number of congruent corresponding angles of a triangle that are needed to determine if two triangles are similar.

2. Distribute the reproducible and explain that students are to follow the guidelines on the reproducible to draw similar triangles.

CLOSURE

Discuss the responses of your students to the question of Step 7 of the reproducible. Students should conclude that the two triangles are similar. Only two corresponding angle measurements are needed to determine if two triangles are similar.

FINDING THE SUM OF THE EXTERIOR ANGLES OF A TRIANGLE

1. Go to http://www.mathwarehouse.com/geometry/triangles/angles/remote-exterior-and-interior-angles-of-a-triangle.php.

2. Scroll down and click on "Make Angle" to create an exterior angle of a triangle. Notice that the exterior angle of a triangle is equal to the sum of the remote interior angles.

3. To verify this principle, scroll down to "Interactive Demonstrations of Remote and Exterior Angles of a Triangle" and drag the vertex.

4. Use the diagram below to answer the questions that follow.

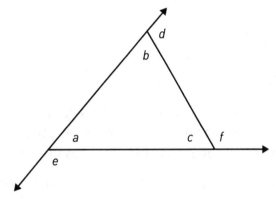

   a. Which angles are exterior angles?

   b. What remote interior angles are paired with each exterior angle?

   c. What is the sum of the exterior angles?

   d. Express the sum in degrees.

1. Draw an angle on your paper. Measure the angle and write its measure in the vertex. Following is an example.

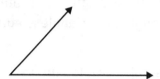

2. Draw another angle so that one of its sides is a common side of the angle you drew in Step 1. Measure this angle and place the measure in the vertex.

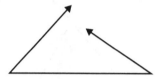

3. Extend the sides of these two angles to form a triangle. (If the sides do not form a triangle, adjust either one or both of the angles you drew. Remember to measure the new angle or angles and place the measure in the vertex or vertices.)

4. On another sheet of paper, draw another angle that is congruent to the angle you drew in Step 1, making its sides a little longer or shorter than the sides of the angle in Step 1. Place the measure of the angle in its vertex.

5. Draw an angle so that one side of the angle you drew in Step 4 is a common side of this new angle and this new angle is congruent to the angle you drew in Step 2. Measure this angle and place the measure in its vertex.

6. Extend the sides of these two angles to form a triangle.

7. Answer the following question: How does the triangle you drew in Step 3 compare with the triangle you drew in Step 6? Explain your answer.

Geometry: 8.G.6

"Understand and apply the Pythagorean Theorem."

6. "Explain a proof of the Pythagorean Theorem and its converse."

BACKGROUND

Pythagoras was a Greek philosopher and mathematician who lived in ancient Greece from about 582 to 500 BC. He is given credit for the theorem that bears his name. Pythagoras discovered that in a right triangle, the square of the length of the hypotenuse, c, is equal to the sum of the squares of the lengths of the legs, a and b. If $\triangle ABC$ is a right triangle, then $c^2 = a^2 + b^2$.

The converse of the Pythagorean Theorem states that if the square of the longest side of a triangle, c, is equal to the length of the sum of the squares of the lengths of the other two sides, a and b, then the triangle is a right triangle. If $c^2 = a^2 + b^2$, then $\triangle ABC$ is a right triangle.

 ## ACTIVITY 1: I CAN EXPLAIN IT

Working in pairs or groups of three, students will individually select a proof of the Pythagorean Theorem and its converse from texts, reference books, or Web sites. They will explain the theorem and its converse to their partner (or partners), who will ask questions, if necessary, to clarify the explanation. The students then switch roles, the first student listening to the explanation of the second student. Students will then write an explanation of their proof and its converse.

MATERIALS

Computers with Internet access. Optional: graph paper; rulers; protractors; scissors; math reference books.

PROCEDURE

1. Explain to your students that they will work together to explain proofs of the Pythagorean Theorem and its converse.

2. Discuss the Pythagorean Theorem and its converse with your students. Note that the theorem and its converse have been proven in several ways by numerous individuals. Many of these proofs are available online as well as in math texts and math reference books. Suggest that students consult math reference books (if available) or do a Web

search for "Proofs of the Pythagorean Theorem." They may also visit the following Web sites:

- http://www.mathisfun.com/pythagoras.html
- www.cut-the-knot.org/pythagoras/index.shtml
- http://www.davis-inc.com/pythagor/index.shtml

3. Instruct your students to select a proof of the theorem and its converse. They are to explain this proof and its converse to their partner (or partners). Partners should help each other clarify their explanations by asking questions or providing suggestions.

4. Encourage your students to use graph paper, rulers, protractors, and scissors if their proof requires sketches or drawings.

5. After students have discussed their proofs and converses with their partners, each student is to write an explanation of his or her proof and its converse.

CLOSURE

Ask volunteers to share their explanations with the class. Display the written proofs.

 ## ACTIVITY 2: THE PYTHAGOREAN PUZZLE

For this activity, students will visit a Web site of the National Library of Virtual Manipulatives to virtually prove the Pythagorean Theorem.

MATERIALS

Computers with Internet access.

PROCEDURE

1. Explain to your students that they will virtually prove the Pythagorean Theorem by completing two puzzles. Each puzzle requires students to fill two congruent regions by rotating and translating congruent triangles and squares.

2. Instruct your students to go to the Web site http://nlvm.usu.edu/en/nav/vlibrary.html. They should click on "Geometry," grades 6–8, scroll down to the Pythagorean Theorem and click on the link. Note that they will see the first puzzle on the screen. Directions for moving and rotating the figure are included on the right-hand side of the screen.

3. Instruct your students to complete the first puzzle by filling each white region with the congruent triangles and square that are below each area. Completing the puzzle correctly illustrates the Pythagorean Theorem. After completing the first puzzle, students should sketch the positions of the triangles and square that they placed in each white region. They should also write an explanation of how their sketch illustrates the Pythagorean Theorem.

4. After your students have finished the first puzzle, they should click on "Go to Puzzle 2," and follow the same procedure as they did in Puzzle 1. Note that the figures students will be using to fill the white region vary from the figures in Puzzle 1.

CLOSURE

Possible sketches for the puzzles are shown below.

Puzzle 1

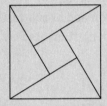 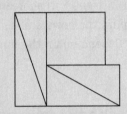

Puzzle 2

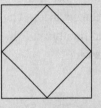 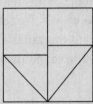

Ask student volunteers to explain how their sketches proved the Pythagorean Theorem. Answers may vary. **A possible answer for Puzzle 1 follows:** The white region on the left has an area of c^2. The white region on the right has an area of $a^2 + b^2$. As each white region is filled with the same figures, $c^2 = a^2 + b^2$. **A possible answer for Puzzle 2 follows:** The two white regions are the same. The white region on the left is filled with four congruent triangles and a light green square whose area is c^2. The white region on the right is filled with the same four congruent triangles and a light blue square whose area is b^2 and a royal blue square whose area is a^2. If the four congruent triangles are removed from each white region, $c^2 = a^2 + b^2$.

Geometry: 8.G.7

"Understand and apply the Pythagorean Theorem."

7. "Apply the Pythagorean Theorem to determine unknown side lengths in right triangles in real-world and mathematical problems in two and three dimensions."

BACKGROUND

To find the unknown side of a right triangle, students should determine if they are to find the length of the hypotenuse, c, which is opposite the right angle, or the length of a side, a or b, which is adjacent to the right angle. Making a sketch of a right triangle with the appropriate labels can help students visualize and understand the lengths they are trying to find. Then they should substitute the appropriate numbers or a letter into the Pythagorean Theorem, $c^2 = a^2 + b^2$. To solve the equation, they should simplify the powers first, then add or subtract to isolate the variable, and finally find the square root. Depending on the numbers, students may have to round their answers.

 **ACTIVITY: APPLYING THE PYTHAGOREAN THEOREM**

This activity focuses on real-world applications of the Pythagorean Theorem. Working in pairs or groups of three, students will find real-world problems that can be solved by using the Pythagorean Theorem, sketch the triangle, and solve problems.

MATERIALS

Math texts; computers with Internet access.

PROCEDURE

1. Instruct your students to find five real-world problems that can be solved using the Pythagorean Theorem. Suggest that they search the Internet, using terms such as "Applications of the Pythagorean Theorem," or consult their math texts. You may encourage them to create problems of their own, based on information they may find or know from their own experiences. Depending on the problem, some answers may be irrational and should be rounded.

2. Offer this example of a problem: A 20-foot fir tree must be staked to prevent the wind from blowing it over. The base of the tree is surrounded by small shrubs arranged in a circle with a 3-foot radius around the tree. To avoid the shrubs, the stake must be set 3 feet from the base of the tree. Ask your students about how much rope is required to support the tree if the rope is tied to the stake and around the trunk of the tree 15 feet from the ground. Ask volunteers to solve the problem, then demonstrate the solution, using the illustration below.

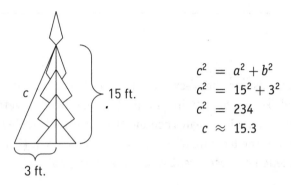

$$c^2 = a^2 + b^2$$
$$c^2 = 15^2 + 3^2$$
$$c^2 = 234$$
$$c \approx 15.3$$

3. Explain that the rope must be at least 15.3 feet to reach from the stake to the tree. (The hypotenuse of the right triangle in the sketch is about 15.3 feet.) Note that this answer is rounded to the nearest tenth. Students may need to round some of their answers. Because additional rope is required to tie around the tree and the stake, a fair estimate for the total length of rope needed to stake the tree is about 18 feet.

4. Explain to your students that after finding their problems, they are to write the problems down, sketch the right triangles, and solve the problems using the Pythagorean Theorem.

CLOSURE

Ask students to share their results. Focus the discussion on how to use the Pythagorean Theorem to solve real-world problems.

Geometry: 8.G.8

"Understand and apply the Pythagorean Theorem."

8. "Apply the Pythagorean Theorem to find the distance between two points in a coordinate system."

BACKGROUND

Students can find the distance between two points in the coordinate plane, such as the distance between (x_1, y_1) and (x_2, y_2), by drawing a right triangle with the vertex of its right angle at (x_2, y_1) and applying the Pythagorean Theorem. The length of the leg parallel to the y-axis is $|y_2 - y_1|$. The length of the leg parallel to the x-axis is $|x_2 - x_1|$. (It is necessary to find the absolute value of the lengths because distance is always positive.)

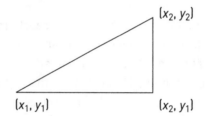

By substituting $|x_2 - x_1|$ for a and $|y_2 - y_1|$ for b in the Pythagorean Theorem, $c^2 = a^2 + b^2$. Students can solve for c, which is equal to $\sqrt{a^2 + b^2}$.

 ACTIVITY: FINDING THE DISTANCE

Students will work at a Web site and, using the Pythagorean Theorem, virtually find the distance between two points on the coordinate plane.

MATERIALS

Computers with Internet access.

1. Instruct your students to go to the Web site http://www.mathwarehouse.com/algebra /distance_formula/interactive-distance-formula.php. (Although this Web site uses the distance formula, students can also find the distance between two points using the Pythagorean Theorem.)

2. Explain that students will use this Web site to find the distance between two points on the coordinate plane by using the Pythagorean Theorem. Using the triangle that appears on the screen, they will drag a vertex to create a right triangle. Note that one of the legs will be parallel to the x-axis and the other leg will be parallel to the y-axis. The lengths of the legs will be calculated for them.

3. Explain that students are to substitute the lengths of the legs for a and b in the Pythagorean Theorem and use the Pythagorean Theorem to find the value of c.

4. Allow your students time to create other right triangles and solve for c. Encourage them to verify their answers.

CLOSURE

Ask your students if they were able to verify all of their answers. In some cases, the answers may not be exact because of rounding. Also, ask students how the distance formula, $d = \sqrt{(x_2 - x_1)^2 + (y_2 - y_1)^2}$, is derived from the Pythagorean Theorem. Students should realize that d is the length of the hypotenuse of a right triangle whose vertex is (x_2, y_1) or (x_1, y_2).

Geometry: 8.G.9

"Solve real-world and mathematical problems involving volume of cylinders, cones, and spheres."

> 9. "Know the formulas for the volumes of cones, cylinders, and spheres and use them to solve real-world and mathematical problems."

BACKGROUND

The formula for finding the volume of some three-dimensional figures follows:

$$\underline{\text{Cone: }} V = \frac{1}{3}Bh \text{ or } V = \frac{1}{3}\pi r^2 h$$

$$\underline{\text{Cylinder: }} V = Bh \text{ or } V = \pi r^2 h$$

$$\underline{\text{Sphere: }} V = \frac{4}{3}\pi r^3$$

For the formulas for the cone and cylinder, B is the area of the circular base, r is the radius of the base, and h is the height of the cone or cylinder. For the sphere, r is the radius of the sphere.

When finding volume, some problems may involve several steps. For example, if students are given the diameter of a cone, they must find the radius first and then substitute this value in the formula.

 ACTIVITY: FINDING THE VOLUME

Students will complete a puzzle by finding the volume of common cones, cylinders, and spheres. Solving the problems will reveal to them a famous Greek mathematician and scientist.

MATERIALS

Reproducible, "Finding the Volume of Cones, Cylinders, and Spheres."

PROCEDURE

1. Explain to your students that we live in a three-dimensional world of height, width, and depth. Examples of prisms, cones, cylinders, and spheres are all around us. Ask your students to offer some examples such as traffic or ice cream cones, pipes (cylinders), and globes (spheres).

2. Review the volume formulas with your students.

3. Distribute copies of the reproducible. Explain that by solving the problems students will discover the mathematician who found that the volume of a sphere is two-thirds the volume of the cylinder that circumscribes the sphere. Note that some problems will require several steps. Also note that all answers are rounded to the nearest whole number and that all units are cubic inches.

CLOSURE

Check students' work.

ANSWERS

The answer is Archimedes: **C.** 29; **M.** 14; **D.** 58; **S.** 382; **H.** 101; **R.** 8; **I.** 5; **E.** 90; **A.** 57. You might want to suggest to your students to research Archimedes to learn more about his accomplishments, especially in plane and solid geometry.

FINDING THE VOLUME OF CONES, CYLINDERS, AND SPHERES

This famous Greek mathematician found that the volume of a sphere is two-thirds the volume of the cylinder that circumscribes the sphere. To find the name of this mathematician, find the volume of each figure (rounded to the nearest cubic inch). Place the letter of the problem in the space above its answer at the bottom of the page.

C. A soda can. It is 6 inches high and has a diameter of 2.5 inches.

M. A cone-shaped ice-cream dish. It is 6 inches high and has a circular top with a 3-inch diameter.

D. A coffee can. The diameter of the top of the can is 3.75 inches. The height of the can is 5.25 inches.

S. A ball. It has a 9-inch diameter.

H. A cylindrical vase. It is 10.5 inches tall and has a radius of 1.75 inches.

R. A sugar cone. It is 5 inches long and has a 2.5-inch diameter.

I. A spice jar. It is 2.25 inches high and the diameter of its base is 1.75 inches.

E. A small bowl (half of a sphere). It has a height of 3.5 inches and a diameter of 7 inches.

A. A cake cone.

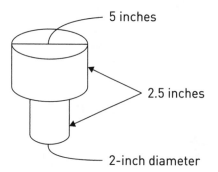

5 inches

2.5 inches

2-inch diameter

___	___	___	___	___	___	___	___	___	___
57	8	29	101	5	14	90	58	90	382

Statistics and Probability: 8.SP.1

"Investigate patterns of association in bivariate data."

1. "Construct and interpret scatter plots for bivariate measurement data to investigate patterns of association between two quantities. Describe patterns such as clustering, outliers, positive or negative association, linear association, and nonlinear association."

BACKGROUND

A scatter plot, also called a scattergram, is a graph of pairs of numbers that is used to analyze the relationship between two quantities. To make a scatter plot, students must take corresponding values from two sets of data and graph the ordered pairs.

The points on a scatter plot may lie on or around an imaginary line, in which case the association is linear.

- If the points on a scatter plot suggest a line with a positive slope, then there is a positive association between the points. (As x increases, y increases.)

- If the points on a scatter plot suggest a line with a negative slope, then there is a negative association between the two points. (As x increases, y decreases.)

If the points do not lie in or around an imaginary line, then the association is nonlinear, or there is no association. In addition to examining the various associations between the data, students may also look for outliers (numerical data that are much smaller or much larger than the rest of the data) and clustering (data that are grouped together).

 ACTIVITY: MAKING A SCATTER PLOT

Working in pairs or groups of three, students will gather data to determine if there is a relationship between the height of an adult and his or her shoe size. They will then make two scatter plots—one for males and one for females—and investigate the patterns of association.

MATERIALS

Graph paper. Optional: a graphing calculator for each student.

PREPARATION

Allow time for students to gather their data. If your students are using a graphing calculator to make their scatter plots, review the keystrokes for entering data in a list, setting the window, and making the scatter plot.

1. Explain that patterns are often found in real-world data. But unless patterns are analyzed, they may provide little useful information. One way to analyze data is to create a scatter plot. Review the vocabulary—clustering, outliers, and associations—and the components of a scatter plot:

    - The title is clear.

    - The values on the *x*- and *y*-axes are labeled.

    - The scale is appropriate.

    - A point is graphed for each piece of data.

2. Explain that, working individually, each student is to ask ten people—five females and five males—their shoe size and height. They should record the information, including each person's name, designating the data obtained from the females with an "F" and the data obtained from the males with an "M."

3. Students will now work with a partner (or partners). Explain that they are to combine their data and create two scatter plots. (Note that in the rare case where two students obtain data from the same person, this data should be used only once.) One scatter plot is to represent the data obtained from the females and the other is to represent the data obtained from the males.

4. Ask your students to identify patterns, especially clustering, outliers, and any associations—positive, negative, linear, or nonlinear.

CLOSURE

Discuss your students' scatter plots. Ask for volunteers to share their scatter plots and note any patterns. Consider questions such as the following: What similarities, if any, were found on the female scatter plots? What similarities, if any, were found on the male scatter plots? Were there any similarities between the female and male scatter plots? What might be the reasons for any similarities or differences?

Statistics and Probability: 8.SP.2

"Investigate patterns of association in bivariate data."

> 2. "Know that straight lines are widely used to model relationships between two quantitative variables. For scatter plots that suggest a linear association, informally fit a straight line, and informally assess the model fit by judging the closeness of the data points to the line."

BACKGROUND

There is often no single line that passes through every point on a scatter plot. A line of best fit (or a trend line) is a line that can be drawn to "fit" the data.

The correlation coefficient is the number r, which indicates how well a set of data can be approximated by the line of best fit. $-1 \leq r \leq 1$.

- A perfect positive correlation is given the value 1.

- A perfect negative correlation is given the value -1.

- If there is no correlation, the value is 0.

 ### ACTIVITY: DRAWING THE LINE OF BEST FIT

Students will virtually create a scatter plot, draw the line of best fit, and compare their line of best fit with a computer-generated line of best fit.

MATERIALS

Computers with Internet access.

PROCEDURE

1. Instruct your students to go to the Web site http://www.shodor.org/interactive /activities/regression/.

2. Explain that they will create a scatter plot by plotting points of their choice. Note that students may enter each ordered pair in the data box beneath the graph and click on "Update Plot," or graph points on the grid by clicking their mouse. In this case, the values of the ordered pairs will appear in the data box.

3. After their scatter plot is completed, students should create a line of best fit by clicking the box "Fit your own line." Remind your students that a line of best fit should follow the line suggested by the points on the scatter plot. Some points will be above the line and some points will be below the line. Students can adjust their line of best fit by checking the dot "Move Your Fit Line" and clicking on the green circles moving the cursor.

4. After students are sure that their line of best fit is accurate, they can check the box "Display line of best fit," and compare this line with theirs. They should note the value of r and the equation of the line of best fit.

5. Suggest that students check the circles "Add Points" and "Remove Points" to see the effect of additional data or the removal of outliers.

6. By clicking on "Reset" students may create another scatter plot.

CLOSURE

Have students write an explanation of how a scatter plot and a line of best fit are related.

Statistics and Probability: 8.SP.3

"Investigate patterns of association in bivariate data."

3. "Use the equation of a linear model to solve problems in the context of bivariate measurement data, interpreting the slope and intercept."

BACKGROUND

A linear equation, $y = mx + b$, can be used to model situations that have a constant rate of change. The graph of the equation is a line. In the equation, m represents the slope (the rate of change) and b stands for the y-intercept (the value when $x = 0$). The equation can be used to solve problems by substituting values for x or y.

Suppose a student deposited $100 in a savings account and put $50 each month into the account without withdrawing any money from the account. This can be represented by the equation $y = \$50x + \$100$, where x is the number of months and y is the total amount in the account. The y-intercept of a graph of this equation is $100, the amount in the account before any monthly savings have been deposited. The slope $50 or $\dfrac{\$50}{1}$ means that $50 is saved every month. The equation can be used to solve problems such as:

- How much money will be in the account after three months? ($250).

- How much money will be in the account at the end of the year? ($700)

- There is $500 in the account, how long has the account been opened? (8 months)

The equation $y = mx + b$ can also be used to solve problems in the context of bivariate measurement data where two measurements are taken from each person or object.

 ACTIVITY: USING LINEAR EQUATIONS TO SOLVE PROBLEMS

Working in pairs or groups of three, students will find examples of bivariate data in a text, reference book, or an online source. They will virtually enter data to display a scatter plot and an equation of the line of best fit. They will interpret the slope, y-intercept, and write and solve two problems that can be solved using the equation.

MATERIALS

Math books (preferably other than the text you are using); science books; other references; computers with Internet access.

PROCEDURE

1. Explain that linear equations can be used to model real-world situations that have a constant slope. Emphasize that the general form of a linear equation is $y = mx + b$.

2. Explain that students are to gather bivariate data from a math book, science book, reference book, or online sources. For example, bivariate values would be the heights and weights of football players on a team. Player 1 is 75 inches tall and weighs 230 pounds; Player 2 is 76 inches tall and weighs 254 pounds; and so on.

3. Instruct students to enter their data virtually at http://www.alcula.com/calculators/statistics/linear-regression/. By clicking on "Submit data," a scatter plot, a line of best fit, and an equation appear.

4. Once students find an equation, they should record it, identify what x and y represent in the context of the data, and note the source; for example, title of the book and page, or the URL of an online source. Students should interpret the slope and y-intercept in the context of their data.

5. Instruct your students to pose two questions that can be solved using their equation and record the answers on the back of their paper.

CLOSURE

Ask for volunteers to present their equation to the class and pose their questions for the class to answer.

Statistics and Probability: 8.SP.4

"Investigate patterns of association in bivariate data."

4. "Understand that patterns of association can also be seen in bivariate categorical data by displaying frequencies and relative frequencies in a two-way table. Construct and interpret a two-way table summarizing data on two categorical variables collected from the same subjects. Use relative frequencies calculated for rows or columns to describe possible association between the two variables."

BACKGROUND

A *two-way frequency table* consists of rows and columns. The entries in the body of the table include joint frequencies. The rows and columns repeat the marginal frequencies or marginal distributions.

Consider the results of a poll of Mrs. Abbot's 30 eighth-grade students who were asked if they liked rock and roll or country music. The results of the poll are summarized in the table below:

		Like Country Music?		
		YES	NO	
Like Rock and Roll Music?	YES	15	1	Total 16
	NO	9	5	Total 14
		Total 24	Total 6	Total 30

- 15 students like both rock and roll and country music.

- 1 likes rock and roll but not country music.

- 9 like country music but not rock and roll.

- 5 like neither.

15, 9, 1, and 5 are the entries in the body of the table.

The marginal row and column frequencies are also shown:

- 16 students like rock and roll.

- 14 students do not like rock and roll.

- 24 students like country music.

- 6 students do not like country music.

A *relative frequency table* expresses each entry, row, and column frequency as a ratio, comparing the frequency of each entry to the total number of people polled. A relative frequency table is shown next.

Like Country Music?

Like Rock and Roll Music?		YES	NO	
	YES	$\frac{15}{30} = 0.5$	$\frac{1}{30} = 0.0\overline{3}$	Total $\frac{16}{30} = 0.5\overline{3}$
	NO	$\frac{9}{30} = 0.3$	$\frac{5}{30} = 0.1\overline{6}$	Total $\frac{14}{30} = 0.4\overline{6}$
		Total $\frac{24}{30} = 0.8$	Total $\frac{6}{30} = 0.2$	Total 1

A *relative frequency of rows table* showing the ratio of each entry to the total of the row that it is in follows.

Like Country Music?

Like Rock and Roll Music?		YES	NO	
	YES	$\frac{15}{16} = 0.9375$	$\frac{1}{16} = 0.0625$	Total $\frac{16}{16} = 1$
	NO	$\frac{9}{14} = 0.6\overline{48571}$	$\frac{5}{14} = 0.3\overline{571428}$	Total $\frac{14}{14} = 1$

A *relative frequency of columns table* showing the ratio of each entry to the total in the column it is in follows.

Like Country Music?

Like Rock and Roll Music?		YES	NO
	YES	$\frac{15}{24} = 0.625$	$\frac{1}{6} = 0.1\overline{6}$
	NO	$\frac{9}{24} = 0.375$	$\frac{5}{6} = 0.8\overline{3}$
		Total $\frac{24}{24} = 1$	Total $\frac{6}{6} = 1$

You can conclude that in this class students who like rock and roll music also like country music. This conclusion can be supported by the following data: 50% of the class likes both rock and roll and country music; about 94% of the students who like rock and roll also like country music; about 63% of the students who like country music also like rock and roll.

 ACTIVITY: ANALYZING TWO-WAY TABLES

Students will work in pairs or groups of three and construct and analyze a two-way table to determine if there is an association between playing sports and being on the honor roll.

MATERIALS

Reproducible, "Sports/Honor Roll Poll."

Prior to the day you plan to assign this activity, distribute a copy of the reproducible to the class. The copy should be passed to all students, who should indicate whether or not they play sports by writing "Yes" or "No," and indicate whether or not they made the school's honor roll by writing "Yes" or "No." (Students' names are not required on the sheet.) After students have finished, collect the sheet. Make enough photocopies of the completed sheet so that you may distribute one copy to each pair or group of students.

PROCEDURE

1. Hand out the completed reproducibles and explain that this is the poll they previously answered.

2. Explain that students will use the data on the sheet to create a frequency table as pictured below.

Have you been named to the honor roll this year?

		YES	NO	
Have you or will you play a sport this year?	YES			Total
	NO			Total
		Total	Total	

3. Explain how to complete the frequency table, a relative frequency table, a relative frequency of rows table, and a relative frequency of columns table. You may wish to provide the examples of these tables that appear in the Background of this activity.

4. Ask your students to use the data from their tables to determine if there is an association between playing sports and being on the school's honor roll. They should support their conclusions with the data.

CLOSURE

Discuss students' findings, particularly their conclusions, which may vary depending on their interpretation of the data.

Have you or will you play a sport this year? Answer "Yes" or "No."	Have you been named to the honor roll this year? Answer "Yes" or "No."

INDEX

Pre-image, 203, 207, 210, 215
Presenting Properties activity, 35–36; background information for, 35; closure for, 36; for expressions and equations, grade 6, standard 3, 35; materials for, 35; procedure for, 36
Prism, 63; volume of right rectangular, 57
Probability, 139–142; approximate, 145–146; compound, 148–150; experimental, 146–147; simulations, 143–144
Probability Investigations reproducible, 141–142
Probability Simulations: background information for, 143; closure, 144; materials for, 143; procedure, 144; for statistics and probability, grade 7, standard 6, 143
Properties, 35–36
Proportion, 84, 86
Proportional relationships, 84–85, 169–170
Proportions Scavenger Hunt activity: background information for, 84; closure, 85; materials for, 84; procedure, 85; for ratios and proportional relationships, grade 7, standard 2, 84
Pyramid, 63
Pythagoras, 227
Pythagorean Puzzle: closure, 229; for geometry, grade 8, standard 6, 228; materials for, 228; procedure, 228–229
Pythagorean Theorem, 227–229; applying, 230–231; converse of, 227; to find distance, 232–233; proving, 228–229

R

Radicals, 160–168
Random sampling, 129–132, 136
Rate, 80; equivalent, 82
Rate of change, finding, 191, 197
Ratio, 80; definition of, 2; understanding concept of, 2–5
Ratios All Around Us activity: background information for, 2; closure, 3; materials for, 2; procedure, 2–3; for ratios and proportional relationships, grade 6, standard 1, 2
Rectangle, formula for finding area of, 127
Rectangular prisms, volume of, 119
Reflection, 201, 203, 205, 210, 215
Relationships, examining, 51–53
Relative frequency table, 243, 244
Rewriting Expressions: background information for, 106; closure, 107; for expressions and equations, grade 7, standard 2, 106; materials for, 106; procedure, 106–107
Right rectangular prism, 57

Right rectangular pyramid, 119
Rotation, 201, 203, 205, 210, 215

S

Same side exterior angles, 219
Same side interior angles, 219
Sample space, 148
Samples, 131; examining, 129–130
Scale drawing, 114–115
Scale factor, 114
Scalene triangles, 116
Scaling Your Classroom activity: background information for, 114; closure, 115; for geometry, grade 7, standard 1, 114; materials for, 114; preparation for, 115; procedure for, 115
Scatter plot, making, 237–238
Scattergram. *See* Scatter plot, making
Scientific notation, 162–164; operations in, 165–167
Shape (statistical), 67–68
Similarity, 201, 205, 210, 215, 219–220; angle-angle, 223–224
Slice of Life with Variables and Expressions, 42–43; background information for, 42; closure, 43; for expressions and equations, grade 6, standard 6, 42; materials for, 42; procedure for, 42–43
Slicing Figures activity: closure, 120; for geometry, grade 7, standard 3, 120; materials for, 120; procedure for, 120
Slope, 169–173
Slope Is the Same activity: closure, 171; for expressions and equations, grade 8, standard 6, 171; materials for, 171; procedure, 171; Student Guide for, 173
Snork's Long Division activity, 12–13; background information for, 12; closure for, 13; materials for, 12; for number system, grade 6, standard 2, 12–13; procedure for, 12–13
Solving Inequalities activity: closure, 113; for expressions and equations, grade 7, standard 4, 112; materials for, 112; procedure, 112–113
Sphere: finding volume of, 236; formula for finding volume of, 234
Sphere, formula for finding volume of, 234
Spinner Experiment activity: background information for, 145; closure, 146; materials for, 145; procedure, 146; for statistics and probability, grade 7, standard 7, 145
Spread (statistical), 67–68
Square, formula for finding area of, 127

Square root, 160–161

Statistical question, 65–68

Statistical Questions Versus Nonstatistical Questions activity: background information for, 65; closure, 66; procedure, 65–66; for statistics and probability, grade 6, standard 1, 65

Statistical variability, 65–71

Statistics, 65

Stimulating Events activity: closure, 150; procedure, 149–150; for statistics and probability, grade 7, standard 8, 149

Stock Market, 98–103

Stock Market Recording Sheet--example, 101–102

Stock Market Worksheet, 103

Student Guide for Dilating Figures reproducible for I've Scrambled My Notes activity and I Found the Image activity, 212

Student Guide for I Have Derived It activity, 174–175

Student Guide for The Slope Is the Same activity, 173

Student Guide for Transforming Figures for Find the Image activity and From Here to There activity and I've Scrambled My Notes activity and I Found the Image activity, 208–209

Students Teaching activity: closure, 182; for expressions and equations, grade 8, standard 8, 181; materials for, 181; procedure, 181–182

Substitution, 39

Summarizing Data activity: background information for, 75; closure, 76; Guidelines for Summarizing Data reproducible for, 77; procedure, 76; for statistics and probability, grade 6, standard 5, 75

Supplementary angles, 124

Surface area, 54

T

Table, constructing, 6

Terms, 104

Test scores, analyzing, 135

Tetrahedron, 63

Theoretical probability, 143–146

This One Does Not Belong reproducible, 97

Three in a Row: answers, 40; background information for, 39; closure, 40; Equations and Inequalities reproducible for, 41; for expressions and equations, grade 6, standard 5, 39; preparation for, 39; procedure, 39–40

Transformations, 201–209; using, 203–204

Transformations activity: background information for, 201; closure, 202; for geometry, grade 8, standard 1, 201; preparation, 202; procedure, 202; Using Transformations reproducible for, 203–204

Translation, 201, 203, 210, 215

Transversals, 222–223

Trapezoid, formula for finding area of, 127

Triangles, 116–117; drawing similar, 226; finding sum of exterior angles of, 221–222, 225; finding sum of interior angles of, 220–221; formula for finding area of, 127; investigating, 118

Two-way frequency table, 243; analyzing, 244–245

U

Unit rate: finding, 80–83; understanding, 4–5

Unit Rate Tic-Tac-Toe activity, 4–5; background information for, 4; closure for, 5; procedure, 4–5; for ratios and proportional relationships, grade 6, standard 2, 4

Using a Function Machine activity: background information for, 183; closure, 184; for functions, grade 8, standard 1, 183; materials for, 183; procedure, 183–184

Using Linear Equations to Solve Problems activity: background information for, 241; closure, 242; materials for, 241; procedure, 242; for statistics and probability, grade 8, standard 3, 241

Using Transformations reproducible for Transformations activity, 203–204

V

Values, 34

Variability, measure of, 69–70, 133, 136

Variables, 42–43; quantitative relationships between dependent and independent, 51–53

Vertical angles, 124, 219

Very Interesting activity: closure, 89; optional materials for, 89; procedure for, 89; for ratios and proportional relationships, grade 7, standard 3, 89

Virtual Classroom activity: background for, 90; closure, 91; materials for, 91; for number system, grade 7, standard 1, 90; procedure, 91

Virtual cube, 119

Virtual Cube activity: background information for, 119; closure, 120; for geometry, grade 7, standard 3, 119; materials for, 119; procedure, 119–120